Space Systems and Sustainability

Joseph N. Pelton

Space Systems and Sustainability

From Asteroids and Solar Storms
to Pandemics and Climate Change

 Springer

Joseph N. Pelton
International Association for the Advancement of Space Safety (IAASS)
International Space University
Arlington, VA, USA

ISBN 978-3-030-75734-2   ISBN 978-3-030-75735-9   (eBook)
https://doi.org/10.1007/978-3-030-75735-9

This Springer imprint is published by the registered company Springer Nature Switzerland AG
The registered company address is: Gewerbestrasse 11, 6330 Cham, Switzerland

This book is dedicated to those that believe the time has come for those on Earth to collaborate in new and more effective ways to save our cosmic home. It is about using space systems and technology to take better care of our ailing Spaceship Earth.

The main purpose of such a global collaboration is to save and sustain civilization from a growing number of existential crises. Some of these threats come from cosmic sources such as asteroids, comets, centaurs, solar storms, and orbital space debris. Then there are also the many and growing crises here on Earth. These include pandemics, natural disasters, global pollution, climate change, nuclear and biochemical weapons, nuclear waste, pathogens, overpopulation, and other "hyperobject" grand challenges discussed in this book.

I dedicate this book to the fledging initiative outlined in Chap. 14, namely the creation of a new international agreement that could ultimately lead to a global sustainability treaty. The goal is to create a unified effort to undertake planetary defense. We need such a global initiative in order to achieve the long-term survival of our species.

Acknowledgment

This book has been a true challenge because of the diverse areas of concern when it comes to long-term human survival in the so-called Anthropocene epoch, which others call the Holocene epoch. We are in a geological time in which human activity is the prime shaper of change to planet Earth. This book thus needed to call upon a wide range of expertise to help readers understand the various crises that the modern world now faces.

I must first start with my thanks to my friend, colleague, and editor, Peter Marshall in England, with whom I have worked on nearly ten different book projects. I truly appreciate his help to format and improve the coherence of this text. I also wish to thank my friend and colleague Dr. Donald Daniel for his expertise in disarmament, nuclear weapons, and nuclear waste materials. He also helped me understand the particular problems with the monitoring of biological weapons. Likewise, my colleague and friend Prof. Scott Madry helped with understanding some of the latest breakthroughs in hyper-spectral sensing and data analytics that are zooming ahead at warp speed.

A special thank you must go to Prof. Ram S. Jakhu, Acting Director of the Institute of Air and Space Law, McGill University, and his colleague, Dr. Karin Vazhapully. They collaborated with me in the writing of Chap. 14, which addresses the creation of a global sustainability treaty. This new international agreement would facilitate the nations of the world to collaborate and identify the world's greatest existential crises, and then to seek unified ways to mitigate these risks.

There are many others from around the world and in many different fields that have assisted in different ways. My sincere appreciation is extended to President Juan de Daumau, Gary Martin, Prof. Su-Yin Tan, and Prof. Chris Welch, all of the International Space University; Chris Johnson of the Secure

World Foundation; Lord Martin Rees, Royal Astronomer of the UK; Matteo Madi of Sirin Orbital Sciences; Olga Sokolova, risk analyst from Zurich, Switzerland; Prof. Steven Freeland of Western Sydney University of Australia; Dr. James Green, Chief Scientist at NASA; Prof. Giovanni Fazio, Harvard University; Michael Potter, Paradigm Inc.; Prof. Madhu Thangavelu , Viterbi School of Engineering, University of Southern California; Ranjana Kaul, space lawyer from New Delhi, India; Dr. Carlos Niderstrasser; Dr. Darren McKnight of Centauri Corporation; Tommaso Sgobba, Executive Director of the IAASS; and Michelle Hanlon, co-founder of For All Moonkind.

Contents

About the Author

Joseph N. Pelton, Ph.D., is the dean emeritus and former chairman of the board of trustees of the International Space University. He is the founder of the Arthur C. Clarke Foundation and the founding president of the Society of Satellite Professionals International—now known as the Space and Satellite Professionals International (SSPI). Dr. Pelton currently serves on the executive board of the International Association for the Advancement of Space Safety. He is the director emeritus of the Space and Advanced Communications Research Institute (SACRI) at George Washington University, where he also served as director of the accelerated masters' program in telecommunications and computers from 1998 to 2004. Previously, he headed the interdisciplinary telecommunications program (ITP) at the University of Colorado-Boulder. Dr. Pelton has also served as president of the International Space Safety Foundation and president of the Global Legal Information Network (GLIN). Further, he served as the founding chairman of the board of The Global Alliance for International Collaboration in Space (GALIX). Earlier in his career he held a number of executive and management positions at COMSAT and INTELSAT, the global satellite organization where he was director of strategic policy.A prolific author, Dr. Pelton has now published over 50 books and over 400 articles, encyclopedia entries, op-ed pieces, and other research publications during his career. He has been speaker on national media in the USA (PBS New Hour, Public Radio's All Things Considered, ABC, and CBS) and internationally on BBC, CBC, and FR-3. He has spoken and testified before Congress and the United Nations, and delivered talks in over 40 countries around the world. His honors include the Sir Arthur Clarke International Achievement Award of the British Interplanetary Society; the Arthur C. Clarke Foundation Award; the ICA Educator's award; the ISCe Excellence in Education Award; and being elected to the International Academy of Astronautics. Most recently, in 2017, he won the Da Vinci Award of the International Association for the Advancement of Space Safety and the Guardian Award of the Lifeboat Foundation.Dr. Pelton is a member of the SSPI Hall of Fame, fellow of the IAASS, and associate fellow of the AIAA. Pelton's *Global Talk* won the Eugene Emme Literature Award of the International Astronautics Association and was nominated for a Pulitzer Prize. His most recent books are: *Preparing for the Next Cyber Revolution, Space 2.0: Revolutionary Advances in the Space Industry, The New Gold Rush: The Riches of Space Beckon, The Handbook of Space Satellites, Global Space Governance: An International Study,* and the second editions of *The Handbook of Satellite Applications* and *The Farthest Shore: A 21st Century Guide to Space.*He received his degrees from the University of Tulsa, New York University, and from Georgetown University, where he received his doctorate.

1

Ten Existential Risks to Life on Earth

We are the intelligent elite among animal life on earth and whatever our mistakes, [Earth] needs us. This may seem an odd statement after all that I have said about the way twentieth century humans became almost a planetary disease organism. But it has taken [Earth] 2.5 billion years to evolve an animal that can think and communicate its thoughts. If we become extinct she has little chance of evolving another.

–James Lovelock, scientist and advocate of the Gaia theory of Earth

Preserving Humanity: Ready, Set, Go!

The *Wired Magazine* devoted its entire April 2020 issue to Earth Day and climate change. This particular edition was emblazoned across the top with the admonition: "We have one Earth—and the technology to save it—Go!"

This book is about this admonition—but with a large caveat. Humans need to be much more careful with the technology we use to save Earth, as inappropriate use of technology can often be a large part of the problem. There is a need to deploy innovative and re-envisioned technology in order to meet a myriad of twenty-first-century challenges. These are real challenges to the long-term survival of the human species in our native habitat, called throughout this book "Spaceship Earth." This is important to emphasize: Earth is indeed a modestly shielded planet, with only a thin protective atmosphere and a puny magnetic cage that protects the world from solar wind and the occasional solar superstorms.

© The Author(s), under exclusive license to Springer Nature Switzerland AG 2021
J. N. Pelton, *Space Systems and Sustainability*, https://doi.org/10.1007/978-3-030-75735-9_1

We have embraced technology to industrialize the world for centuries now. This has been done without adopting protective practices or safeguards for the environment. Too often, technology is seen as a way to increase economic throughput and strengthen corporate profits. This book suggests that there is another and more important role for technology: to create new ways to sustain life on Earth. Profits are fine. Continuation of the species, however, needs to be ranked just a smidgen higher on humanity's value scale.

The willingness of humans and governmental and corporate entities to spend more money and resources to save Earth as a viable biosphere is the great ethical question of contemporary times. Humans now officially live in what is known as the "Anthropocene Epoch." The geologists of the world have decided on this new designation only in the past decade. Anthropomorphic beings—that is, humans—for better or more likely for worse, seem to have pushed Mother Nature out of the way and now dictate the future of the third rock from the Sun.

For millennia, humans relied on their skills as hunters and gathers, but today, we are dependent on increasingly complex technology often separated from the world of nature. This separation has provided many benefits and created wealth, but it has also contributed to new types of crises. Our dependence on essential technology also makes modern society more vulnerable in the event our infrastructure is disabled or destroyed.

There is a need to urgently deploy what might be called sustainable and nature-friendly technology to compensate for the negative impacts that humanity is making on the world through unrestrained growth and environmental excess. This book sets forth ten challenges that humanity now faces as a civilization. These challenges are examined in some depth, along with the needed new technologies that may help solve them. An effective response requires more global cooperation, interdisciplinary skills, and a better overview and understanding of how many of these threats are interrelated. Shifts in policy, regulation, and law are essential to preserving a livable world. The bottom line is that new technology, policies, laws, and regulation, and significant amounts of money, time, and resources, need to be invested. And indeed, some of the technology developed to "save" Spaceship Earth might create their own problems, giving rise to concerns about personal privacy or perhaps runaway space debris, to name only two. Technology is clearly not a cure-all, but rather a tool to be used carefully and wisely.

Slow and lackadaisical reaction by the general public and politicians to these great challenges calls to mind the stock-in-trade joke of radio and television comedian Jack Benny from the 1940s through the 1980s. More than once, Benny is confronted by a robber. As the robber points a gun menacingly and demands "Your money or your life?" Jack Benny just stands there and does not respond. It seems as if minutes go by, and the robber again threatens him urgently at gunpoint: "Listen mac this gun is loaded. I said your money or your LIFE—what is it going to be?" Jack Benny replies, "I'm thinking, I'm thinking."

But global society is not thinking. Political and business leaders are not always considering the needs of everyone on the planet. Often, those in "technically agile" and economically advanced countries are unconsciously choosing their money and luxury lifestyles over the future of life on the planet. Many times the push for corporate profits tends to outweigh consideration of future children, grandchildren, and great grandchildren. In the case of the Jack Benny joke, he is facing an immediate threat but still trying to keep his money even in the face of mortal danger. In the case of most of the challenges that humanity faces today, the mortal threat is perhaps many decades into the future or even longer. Humans do not deal well with longer term risks. It is not easy for the average citizen or a politician to decide how to prioritize spending of resources. How does one choose to spend money wisely? How does one decide to act? Is it best to address a real problem today, such as a pothole in the street? Or is it best to be prepared for a potential cataclysmic event that may hit Earth years from now? In short, humans and dinosaurs often have an equal capacity to think strategically.

"Hyperobject" Threats to Human Civilization

Environmentalist Tim Morton invented a useful term that applies to major long-term societal threats. In Morton's view, such challenges will ultimately be devastating, but they are still likely a long way off and are very complex in a scientific or technological way. He coined the term "hyperobject" to refer to such long-term and very difficult questions. According to Morton, a hyperobject is something so gigantic in its impact, so long-lasting over a substantial period of time, and so complex and difficult to address, that most people simply dismiss it. Let other people face hyperobject challenges. This might perhaps be a scientist in a laboratory somewhere, an astronomer in an observatory, or a meteorologist taking measurements. Often, it is a politician who says, let elected officials that come after me face this problem—the repair of potholes come first. The key is to be able to place the blame elsewhere if something goes very, very wrong. For many leaders, the key is to be well out of office when the s—t hits the fan.

This book suggests that it is time to understand that in the Anthropocene Epoch, the world is changing and existential risks are mounting. Existential risks, or even true threats are problems that impact humanity and all life on Earth, possibly to the catastrophic point of extinction. Significant changes due to massive urbanization, population growth, and reliance on modern infrastructure mean that the challenges once small and remote are now increasingly large and the likelihood of devastating impact greater. As the world becomes more technologically dependent, the fragility of the modern world becomes more real. It is an irony of the time that new technology can make modern society safer and more comfortable, at the same time that it makes society more vulnerable.

Fortunately, many of the newest space systems and associated data analytic capabilities can help identify the most pressing risks and also the best means to address them. New and promising developments are coming from the world of space and satellite systems, information technology, cybersecurity, and environmental sciences. But solutions are not just technological or scientific. In many cases, the answer must involve a change in governmental legislation. It also requires new, proactive policies. One of the solutions offered here is the adoption of a new International Agreement that ultimately leads to a Global Sustainability Treaty.

Don't worry—this is not a proposal to create a new international agency. The purpose of this treaty, or international agreement, is largely to knock down disciplinary silos that keep apart useful information and applications in a world that is overly specialized and does not share information wisely or well. The object of such a treaty would be to have more effective information sharing across disciplines and to work nicely together. It would encourage various existing international and national agencies to work on longer term problems on a wider scale.

Just what are the crises we are facing? Future chapters will reveal their nature, along with the new types of protective systems and technologies that might deployed against them. These challenges to humanity include pandemics and climate change. But there are also eight other challenges to consider. Some are cosmic in nature, including massive solar storms, coronal mass ejections, asteroid strikes, and comets. There are also natural disasters, which in many ways are now more dangerous than ever before and will systematically kill more people and destroy more infrastructure. Still others include modern destructive capabilities, such as overpopulation, nuclear and biochemical weapons, species extinction, and global pollution. Indeed, we even consider artificial intelligence (AI) and various forms of cyberattacks.

The next 80 years, by the start of the twenty-second century, is the critical time to transform to a sustainable society. There are many reforms that are required along the way. There is an urgent need for more worldwide sharing of scientific, technological, and other information to achieve sustainability. This new level of sharing comes from a much broader range of fields than in the past. Some of the key fields are agriculture, banking, finance and taxation, cybernetics, disaster recovery, energy systems, environmental studies, medical research, infectious diseases and pandemics, meteorology, water supply, atmospheric studies, nuclear and biochemical weapons disarmament, space systems (including especially Earth observation systems, precise navigation, positioning, timing, telecommunications, and IT services), transportation and shipping, and much more. There are breakthroughs in many technical applications not even dreamed of just a few years ago. The key is to begin prioritizing them for global sustainability over all other values. We might call this the Jack Benny rule: You can't spend your money if you are dead.

These new capabilities can help provide much greater control from pandemics to pollution cleanup. They can help us monitor natural disasters and nuclear and biochemical weapons. There are new space systems that can provide instant (three times a second) high-resolution imaging of the entire world in near-real time. There are space communications systems that can provide broadband links to any point on the planet, and others that can provide precision navigation, positioning, or timing services. Much of the technology is already here, but it needs to be deployed in the efforts of sustainability. The potential impact of new space applications could turn out to be as significant as the invention of the wheel, farming, electricity, or even spandex.

What follows is a plausible listing of modern crises that human civilization will face in the twenty-first century, as well as a brief introduction to the technology and policies that might help millions of people—quite possibly within the lifetime of anyone reading this book.

Pandemics

The 2020 COVID-19 pandemic calls to mind the "Spanish Flu" of 1917–1918. Incidentally, it is called that because the brave Spanish government was the first to acknowledge this pandemic, not because it started in Spain. If the end result of the COVID-19 virus was to create a stronger global defense against future pandemics and also to address other planetary threats seriously, then it will have served at least one noble purpose. Those that died of this disease will not have done so in vain. The increased linkages of medical research centers, hospitals, and governmental agencies might prove to be one of the best things to happen to humanity in decades.

A shift to the latest in space technology information processing and AI could help detect and contain future pandemic outbreaks with much greater speed. New space systems and data analytics can help monitor native habitats that are known to give rise to zoonotic diseases, that is, diseases that jump from animal hosts to humans. Certainly, new global communication systems can share information much more efficiently with regard to therapeutics and vaccines. Remote-sensing satellites linked to instant data-analytic systems could confirm outbreaks of infectious diseases and pinpoint the location of those infected in a matter of minutes and hours, rather than days or weeks. The right combination of sensors in space and on the ground could signal a pandemic even before a government officially notifies the World Health Organization that there is a problem.

Global coordination of pandemic responses can be greatly improved via broadband networking that is available across the globe. Leaders and medical professionals across these systems can share key information on health practices and medical treatment. These areas will be addressed more fully in Chap. 2.

As we've seen, most political leaders find ways to respond to disasters, but this is typically after the fact and after great damage has been done. There is little or none by way of proactive disaster prevention or mitigation programs in most countries. This may be particularly true with regard to pandemics. Always, it is a matter of trying to lock the barn door after the livestock has escaped.

Sorting out what should now be spent to prevent pandemics is a bit difficult to estimate. Trillions of dollars are now spent on medical care and research around the world. Yet of all the money spent on medical care and research on new medications, hundreds of times less than that amount is devoted to the detection and prevention of pandemics. It is well documented that the unit devoted to pandemic prevention and response established in the White House by the Obama administration was abolished by the Trump administration. The answer to this ill-advised action was something like: "We will recreate such a unit if a need for it arises." The irony of that statement today is overwhelming.

But other governments and organizations are seldom proactive in the prevention of pandemics too. Even though the World Health Organization (WHO) and the U.S. Center for Disease Control (CDC) have traditionally viewed their prime mission as the containment of infectious diseases, their response was clearly lacking. As will be explored in the following chapter, "an ounce of prevention is now worth many pounds of cure." If a billion dollars more had been spent on new technology, alert systems, and pandemic-related protocols, this might have saved many billions more in business, governmental, and personal losses.

The point in Chap. 2 is not to condemn the lack of planning in the past, but to lay out a blueprint of action that involves more effective use of information and detection technology along with proactive programs to prevent a repeat of the COVID-19 disaster anytime in the future.

The Threat of Nuclear and Biochemical Weapons

COVID-19 *could* have been a human-engineered biological weapon released from a weapons lab—but it most certainly wasn't. Conspiracy theorists might have found reason to spread such a rumor, but it is well understood that the virus was most likely spread from either a bat or a monkey in the vicinity of Wuhan, China. Yet, the threat of a biochemical attack or even a nuclear war unfortunately remains a real one in today's world.

If we had seen, for instance, the release of a small pox agent, this would have made the current pandemic seem like a walk in the park. And atomic weapons remain an enormous and legitimate concern, even though they have not killed people in several decades. Legitimate efforts have been made to reduce the stockpile of over 13,000 nuclear weapons, but with little success in recent years. In contrast to the perhaps $20 trillion spent on armaments over the last decade, only a trifling amount has been spent on peace. Today, there are new armament initiatives in the United States, Russia, China, North Korea, and possibly other countries. The threat of mass destruction, as monitored by the Union of Concerned Scientists, remains quite real (Fig. 1.1). There is no truly effective system for containing the spread and possible use of nuclear and biochemical weapons. International agreements on this issue need to be strengthened. Technical systems to monitor and detect their use need to be enhanced. Today, there are monitoring systems on GPS satellite to detect nuclear device explosions and asteroid strikes, but these can only report on such events after the fact.

New initiatives must seek disarmament and monitoring and verification systems. This thinking must be considered a high priority in any rational future of humankind. As these efforts move forward, it should perhaps be recognized that biochemical weapons may be a bigger threat to humanity in the twenty-first century than nuclear weapons. The efforts of rare political leaders, as exemplified by Senator Samuel Nunn in his nuclear decommissioning campaign and others of his ilk around the world, must be initiated with renewed vigor.

Fig. 1.1 The sharp teeth of nuclear weapons. (Cartoon courtesy of the Union of Concerned Scientists)

Potentially Hazardous Asteroids and Bolides

It is little known that even a 30- to 40-meter asteroid could potentially destroy the entire San Francisco Bay area, including Silicon Valley, or megacities such as Beijing, Mumbai, London, Cairo, Lagos, and Mexico City. Yet NASA, under U.S. Congressional directives, was initially told to scan the skies for asteroids that were at least 1 kilometer in diameter.

After some thought and through the leadership of Congressman George Brown, this asteroid survey guidance was reduced down to asteroids at least 140 meters in diameter. This revised figure may seem to be a more logical search strategy. Yet, astrophysicists and astronauts such as Rusty Schweickart and Ed Lu of the B612 Foundation have said not so fast! The Tunguska asteroid that struck Siberia in 1908 was only a 40-meter rock, yet still caused incredible damage to the surrounding area. Further space rocks of the type that caused the Tunguska event are at least a hundred times more common than a 1-kilometer asteroid and tens of times more common than even 140-meter asteroids. The amount of money we devote to detecting world-ending objects and assorted city-killer rocks is less than 0.01% of what we spend on defense. Again and again, the rationale of how humans spend money comes into question.

We now have the technology to detect potentially hazardous asteroids with much greater precision. NASA and other space agencies should be finding asteroid threats above 30 meters in diameter. In addition, there are a host of possible technologies that could be used to detect and divert asteroid threats from creating cataclysmic harm to planet Earth. These need to be funded and developed as a priority. Only recently, a killer asteroid flew near Earth, reminding us that deadly space threats are indeed quite real [1].

Comets and Other Space Dangers

Comets are an even bigger problem than asteroids when it comes to potential extreme danger. This is because comets travel even faster and generally are large enough to do tremendous damage. Additionally, they are in some ways harder to detect. The more than 20 pieces of the comet Shoemaker-Levy 9 that blasted into Jupiter in July 1994 would have ended life as we know it on Earth. Many of these pieces were much greater than the 6-kilometer asteroid that killed off the dinosaurs some 65 million years ago, along with perhaps 80% of all the species on the planet. Techniques developed to address hazards from asteroids might be adapted to comets as well.

Comets, antimatter, and manmade perils such as space debris are discussed in Chap. 7. They are obscure to most of the public, but these could represent true challenges to humanity. Orbital debris, in particular, is a danger to many of types of space systems on which humanity now relies. It is also perhaps the most directly addressable space threat that could be diminished through active debris removal and stringent debris mitigation procedures.

Cosmic threats may seem exotic and unlikely to most people, but they are very real and potentially very dangerous as well. The first step is to develop new forms of space situational awareness, and the second step is to develop better planetary defense capabilities.

Solar Storms and Electromagnetic Pulses

The biggest cosmic hazard that is most likely to occur is a *coronal mass ejection* (CME). These CMEs could contain quintillions of ions traveling at millions of mph (or km/hr), which could wipe out the world's electronic power systems. This same event could also destroy the controls of thousands of pipelines, shut down the Internet, and much more. A Lloyds of London study estimated that a big CME strike could cause close to some 3 *trillion* of damages to the North American grid alone.

The best current estimate is that a massive solar stream of ions typically occurs about every 150 years. The last really big one, called the Carrington Event, came in 1859. This event zapped Earth big time. Telegraph offices caught on fire when ions flew off the wires and onto the paper in these buildings. The Northern Lights moved so far south that they could be seen in Hawaii and Cuba. Records in China seem to show that a similar event occurred there in the early 1700s. We are likely overdue for something as violent and damaging. The CME Halloween occurrence of 2003 in Scandinavia and the similar one that crippled electrical grids from Chicago to Montreal in 1989 were much less violent, yet prove that this phenomenon is a true and tangible danger.

Technologies such as Faraday cages, heavy-duty circuit breakers, and more can be used to defend against a massive CME. But more research is needed to defend vital electronic infrastructure against a giant solar storm. NASA and other space agencies need to look at current changes to the Earth's natural electromagnetic shielding against solar storms. The shift of the magnetic poles that satellite systems rely on has been detected by both NASA and ESA satellites, meaning our natural protective shielding is being systematically degraded. Research scientists and the author of this book have suggested that a space-based shielding could be developed for both the Earth and Mars to lessen the effects of solar wind and solar storms in the form of CMEs to protect against these types of cosmic dangers.

Electromagnetic Pulse (EMP) Events

Most people think of nuclear events that occur in warfare to be in the form of bombs, like those that fell on Hiroshima and Nagasaki, Japan, in August 1945. Yet, we now know that a nuclear device that explodes in the upper atmosphere can create an electromagnetic pulse (EMP). Such an EMP would not destroy buildings or kill people, but it might have enormous destructive impact on the country over which it is exploded. It might knock out most of the computers. It might destroy yottabytes of data stored all over the country and destroy banking and savings records, social security information and medical records, utility information, and more. It might destroy the so-called SCADA devices.

Most people do not know about SCADA networks, yet millions of them control power distribution, elevator systems, traffic lights, water and sewage pumps, pipelines, and utilities. Almost everything that is automated works under their control. Today, these systems are often being retrofitted with the *Internet of Things* (IoT), or addressable smart devices. This transition is happening so fast that the entire world is becoming the Internet of Everything. A high-power EMP, which might be generated by a high-altitude nuclear explosion or come from a powerful solar storm in the form of a CME, would likely knock out almost anything that is electronic. People and buildings would remain, but the economy and society would shut down. Essentially every electronic object in range of the ion blast would be dead. Zapped! Caput!

It would be difficult to distinguish quickly if the EMP catastrophe was triggered by a deliberate attack via a nuclear explosion in the atmosphere or rather by a large solar event in the form of a coronal mass ejection (CME). As noted, smaller solar storms do occur and these tend to cause more localized effects. The Montreal Event of 1989, however, took out transformers and significant electrical power from Chicago to Montreal. Figure 1.2 shows a zapped transformer in Chicago before and after that event.

PJM Public Service
Step Up Transformer

Severe internal damage caused by
the space storm of 13 March, 1989

Fig. 1.2   The damage from a solar CME event that occurred in Chicago in 1989

New programs are needed to provide electromagnetic shielding, high-performance circuit breakers, decentralization and sequestration of urban power supplies, better protection of satellites, pipeline systems, and Internet synchronization processes. Solar space shields might protect against solar storms, but not high-altitude nuclear explosions.

Population Growth and a Disposable Economy

The current *disposable economy*—as opposed to a circular or sustainable economy—is eating up the Earth's natural resources at a rapid clip. It is a world that is growing out of control. In 1800, there were 800 million people. In 1900, there were 1.8 billion people. In 2000, there were nearly 7 billion people. By 2100, there could be as many as 12 billion people. Try to imagine what the world will be like in 2200, 2300, 2400.

A world population that will reach perhaps 12 billion by 2100 is going through a growing list of rare Earth metals and exponentially consuming large amounts of food. A lack of food and water leads to mass relocation, which will prompt territorial wars. The rise of population leads to urbanization, destruction of trees and rainforests, and various forms of pollution to the atmosphere, the rivers and oceans, and the lands.

Spaceship Earth has only so many resources. When these run out, humanity and all living things, except perhaps some bugs and smaller biota, are going to be in big trouble.

It can be easy to see many of the megathreats we face as disjointed from each other. The relationships between pollution, lack of education and health care, vulnerability from huge rocks from the sky, food supplies, supply changes, and more are not always clear. Yet many of our greatest threats are deeply interrelated. In two centuries, we have gone from a world that was less than 20% urban to over 50% urban, and we may hit 80% urban by 2100. Overcrowding, overurbanization, overpopulation, and overconsumption are all part of a human megathreat that is unfolding in slow motion. Yet, this gradual bomb has a force over the longer term that is as great as many nuclear devices.

Our technical agility has allowed this unprecedented growth of human population in just two centuries. Yet this agility has made us all the more fragile. This book, if it achieves no other purpose, will help society at large question what we mean by "progress." There already exist many key technologies that may help us cope with population growth. Satellite capabilities that range from ocean, atmospheric, weather, and climate change monitoring to remote sensing and inventorying of vital supplies will be increasingly vital. These space systems plus enhanced data analytics will track key supplies of water, crops, animal life, trees, forest fires, plant disease, and severe storm systems.

Climate Change and Natural Disasters

Most people think that they understand the term *climate change*. It means that global warming is occurring and the seas are rising. They even might understand that so-called greenhouse gases are trapping heat in the atmosphere. Heat is not escaping into outer space like it used to, and the trapped gas is warming the Earth. But climate change is far more than warmer temperatures. The trapped heat is strengthening storms, typhoons, hurricanes, and tornadoes. Desertification is shrinking arable land. Massive amounts of water once held in underground rivers and aquifers are drying up. Rainfall patterns are changing, with areas of drought getting more arid and wetter areas in some cases getting more rainfall. With more severe storm systems, there are more and more lightning strikes. Significantly more lightning strikes are now resulting in many more forest fires.

Pollution of the oceans is raising the acidity of the water. Oil spills are drifting to the Polar Regions and changing the albedo of the icecaps, partially melting the ice much more quickly. Once the pure water in the ice melts, it become saltwater and does not re-freeze. The warming trend now threatens to thaw the massive frozen peat fields of Siberia and other Northern regions. This thawing peat releases methane, which traps heat more efficiently than carbon dioxide.

It is all of these changes together that result in less arable land for agriculture, which is already being reduced by urbanization and population growth. Rainforests are being depleted to create more farms and provide more food to ever-expanding cities. Millions of people are leaving arid regions to migrate to better climes. These self-reinforcing processes threaten the livability of the planet. All efforts to reduce the release of carbon-based greenhouse gases are offset by net population growth that is still in the 2.5–4% range in some countries. Nigeria, for instance, is set to overtake the United States as the third most populous country at current growth rates.

Perhaps, only modest adjustments to governmental budgets could make a big difference. The range of steps that could be taken to slow climate change could be vast and unconventional, such as to start building solar shields to reduce solar heating of the Earth or to darken clouds and make them more reflective. Dozens of lesser actions could be taken as well, as discussed in Chap. 9. There is a huge amount of useful information now written on climate change. The focus of this chapter will thus be on new space and IT systems.

Nuclear Waste, Industrial Poisons, and Pathogens

There is growing concern about the safe disposal of radioactive waste materials from nuclear power plants, industrial poisons that can pollute water supplies, and the creation of pathogens that can lead to infection, illness, and even genetic damage. Today, even the disposal of plastic is becoming a major issue. Microplastic poisoning of fish and ultimately humans is now recognized as a serious problem. Countries in Asia are banning the import of plastic waste. Many materials such as asbestos, arsenic poisons in flooring materials, and waste products are becoming a serious problem.

 We must develop new processes to cope with the disposal of everything from radioactive materials to Styrofoam, from metallic poisons to various pathogens. The idea of shipping waste to developing countries is no longer an acceptable answer. New technology to process this waste material will become a major challenge of our time.

Artificial Intelligence, Cyber Attacks, and Digital Defense

Virtually all of the issues addressed in this book involve risks and threats that have been identified for some time but are now being raised to new levels of concern. This heightened response is largely because the magnitude of the problem is increasing with more people; larger cities; faster throughput of industrial cycles that produce more waste, pollution, acidity, poison, or radiation; scarcity of resources; vulnerability of vital infrastructure; or some combination of these things.

A new risk factor comes from the rise of AI, super-automation, cybernetics, and criticality of information and data analytics. In modern society, information is power and the index of wealth. The ability to access, deploy, attack, manipulate, and steal information is increasingly the pathway to some form of power. The rising power and impact of information is redefining work, restructuring wealth, and reshaping many aspects of society. This includes questions of personal privacy, concepts of work, compensation systems, education and training, income taxation, government, regulation, legislation, and international relations.

Existential Threats to Humanity Are Increasingly Real

The above existential threats and risks to human survival are real. In most cases, they are only growing. Now is the time to develop and deploy the right technology to lower these threat levels. Some of the technologies are as simple as new systems of planning and development. These might include decentralizing urban centers and creating telecities instead of mega-cities. Other solutions might involve new systems of taxation or cleanup. We could do a great deal more to provide improved warning systems and other mitigative strategies for weather-related events and natural disasters. We could get much better returns in certain technologies than we currently do on government investments that often focus on "tanks, toys, trinkets, and trifles."

Bad news, predictions of doom and gloom, and widespread angst from political and business leaders as well as the popular news media do not play well with the public. Downer news tends to kill morale. Ignoring problems, putting off maintenance, and attacking the most urgent tasks of the day are always easier. All longer term issues are easily deferred. Let the politicians next term worry about solar storms, dangerous rocks from orbit, overcrowding of cities, nuclear disarmament, or any other problem. If this is true of politicians, it is doubly true of corporate executives, whose main job is to worry about the next quarterly profit report.

The Vulnerabilities of the Technological Human

The technological human needs to set new priorities and goals. This means elevating survival and sustainability to make these the top priority within the global economy. This represents a fundamental shift. It means downgrading the priority of profit in a free-enterprise world in favor of a new priority of making survival technologies and systems. There is irony in the fact that such priorities, if done well and with conviction, could actually fuel economic growth and create a surge of employment opportunities. This book suggests that green, sustainable, and proactive programs could lead to prosperity, new jobs, and economic expansion. It is particularly true of countries with limited population expansion. China, which has curtailed its demographic growth and achieved unprecedented wealth expansion, is a strong indicator of how slow or zero-population growth can be a road to riches in the post-postindustrial world.

A key theme of this book is how to address the problem of global employment in a world on the verge of super-automation. Indeed, new nature-friendly technology can also fix problems of city infrastructure, mounting problems associated with income taxation, and other issues in a world where technological change is moving much faster than the processes of governmental officials, politicians, and businesses.

One of the true weaknesses of decision-making in a modern democratic country is its inherently reactive processes. Linear responses within a world that is logarithmically accelerating are inflicting huge damage on the world's environment, economy, and employment systems. We need to start fixing the world and doing it much, much faster. Again, ready, start, go!

The Conventional Approaches Are No Longer Working

The editors of *Wired Magazine* in their climate change issue mentioned at the start of this chapter were able to find bright, young technologists who conceived of innovative ways of rescuing Spaceship Earth. There were cars that ran on compressed air. There was improved concrete through better chemistry. There were more efficient wind turbines and solar electric power systems. Ideas were trumpeted about bigger and better battery systems and ecological solutions to food supplies via meat substitutes. *Wired Magazine* responded to grand technical challenges in the twenty-first century with more environmentally friendly technologies championed by people with clever ideas.

This commendable approach, however, misses much of the point. Not only must we create new technology, but we must also reboot economic goals, revise pricing systems and market dynamics, re-envision employment processes, and change the R&D and implementation processes for new technology. Innovative, environmental-friendly technology alone is just a Band-Aid [2].

The Western democratic world of free enterprise still marches to the tune elucidated by Milton Friedman a half century ago in his *New York Times Magazine* essay "The Social Responsibility of Business Is to Increase Its Profits." This article, published on September 13, 1970, emphasized that a corporation was simply an "artificial instrument" of its collective owners. As such, corporate executives should only carry out the interest of its owners.

This goal, he indicated, will "generally be to make as much money as possible…" He dismissed talk of corporations having social responsibilities such as "providing employment, eliminating discrimination, avoiding pollution, and whatever else may be the catchwords of the contemporary crop of reformers." Corporate social responsibility was, in his opinion, pure bunk. He stated such talk was an attack on the rights of corporate owners and was "preaching pure and unadulterated socialism" [3].

Some would say that Friedman was stating truths about a capitalist society that date back to the times of economist Adam Smith, author of *The Wealth of Nations*. But this is not true. Indeed, some claim that Adam Smith was actually the father of Corporate Social Responsibility (CSR). He contended even before he wrote *The Wealth of Nations* that free-enterprise corporations had a social responsibility to society at large and to its employees. His earlier book was actually entitled *The Theory of Moral Sentiments.* The very start of this book states that corporate activities must consider "the fortunes of others and render their happiness necessary" [4].

Robert Wood Johnson, the top executive of Johnson & Johnson just before its initial public offering of J&J stock way back in 1943, acknowledged his own view of corporate social responsibility. He noted the responsibilities of his corporation were first to "patients, doctors and nurses, mothers and fathers and others who use our products and services, then to our customers and business partners, our employees and our communities, and finally to our shareholders" [5].

The question of our time is how to define social responsibility and respond to the needs of the twenty-first century. This is a time when so many issues of survival are now apparent. If corporate titans do not realize the need for change, the young people of the world certainly do.

Governmental Agencies and Vital Industries

Here are some fundamental questions of modern governmental policy:

1. Should the space agencies of the world, such as NASA, ESA, JAXA, ISRO, Roscosmos, and the Chinese National Space Agency solely prioritize exploring the solar system, space sciences, and applications, or give prime attention to humanity on Earth and protecting vital infrastructure from space threats such as asteroids, comets, and massive solar storms?

2. Should governmental agencies concerned with the environment, transportation, and energy work to promote existing corporate entities, or should they concentrate on implementing green technologies?
3. Should medical research and drug supply companies re-orient themselves to combat world pandemics such as COVID-19 and prioritize vaccine development and production capabilities?

Grand Challenges

Back in 2008, the U.S. National Academy of Engineering convened a panel of 18 world famous engineers and scientists. The effort was chaired by Larry Page, cofounder of Google, and included former Secretary of Defense William Perry, Artificial Intelligence wizard Raymond Kurzweil, inventor of "SIRI" on your smartphone, and other luminaries. The 14 "Grand Challenges" they came up with were as follows:

1. Make solar energy economical
2. Provide energy from fusion
3. Develop carbon sequestration methods
4. Manage the nitrogen cycle
5. Provide access to clean water
6. Restore and improve urban infrastructure
7. Advance health informatics
8. Engineer better medicine
9. Reverse-engineer the brain
10. Prevent nuclear terror
11. Secure cyberspace
12. Enhanced virtual reality
13. Advance personalized learning
14. Engineer the tools of scientific discovery [6]

This was a thoughtful list, but one devised by smart engineers adhering to an old worldview of progress. It was a future that U.S. hi-tech corporations and governmental agencies knew how to pursue well and efficiently. It was a future that the U.S. military-industrial complex was geared for, invented during World War II some three-quarters of a century ago. These very erudite and capable minds thought of cleaner energy, more efficient education, better health, better engineering, and new and improved scientific process—all commendable goals. They started with the most promising technologies on the horizon—not with the existential needs of a world with a population that is growing too fast. They were not asked to face the problem of having a spaceship on which the passengers are eating up its resources faster than they can be replenished. They did not consider the needs of a world that will be over 80% urban and may become overly dependent on vulnerable hi-tech facilities and infrastructure.

Nowhere in their grand challenge technologies were there tools to defend Earth and its modern infrastructure against massive solar storms or asteroid or comet strikes. Nowhere were thoughts given to new sources of employment or resource gaps—gaps that are widening as our population continues to grow. Nowhere was there a careful consideration of the current disposable world economy. Nowhere was there consideration of such matters as rapidly rising ocean acidity, which might destroy the ocean-based algae and plankton that produce over half of the world's oxygen and dispose of carbon-based greenhouse gases.

Planning a better approach to the future must always be considered on a multidisciplinary basis. It requires expertise that cuts broadly across many fields so that all the right questions are asked. This book is about finding these questions and some of the technology that can help solve them. The point is to suggest that the next grand challenge effort should start with human survival as the prime goal and not a byproduct of technical advancements.

Public Policy and Basic Survival Questions

There is a great deal that needs to be accomplished to sustain the lives of the grandchildren and great grandchildren that will live on Spaceship Earth 80 years hence. The good news is that most of the needed technologies on a grand survival list are probably quite feasible. Most could likely be developed within a decade. Many of the needs are actually new systems and economic or taxation policies. The bad news is that it is hard to turn around global public policy without many years of effort. New policy and key changes must first penetrate the highest levels of political and business leadership. It is perhaps a thousand times more difficult to achieve this than to develop the actual technology and the methods of systemic changes. Saving humanity from its own economic and industrial successes is much harder than turning around a thousand aircraft carriers.

A technological society with sustainability as its priority will require much more by way of policy, regulatory, and legal changes than it requires new gadgets and software in space and on the ground. Ultimately, what is needed is one basic shift. This can be expressed as a global shift from profit to survival. It is a change from a disposable world economy to a sustainable or circular economy. We must acknowledge that we are inhabiting a 6-sextillion-ton spaceship with limited resources, sailing through a cosmic sea rife with dangers deadlier than icebergs or dragons. This is a spaceship with an ever-growing number of passengers aboard, yet with a shrinking amount of supplies to eat. It is very simply the time to start living within the available means of planet Earth.

The Warnings of Jacques Ellul

The French philosopher Jacques Ellul wrote a book in 1954 entitled *Technological Society*, in which he expressed concern about technology not harmonized with nature. He warned that such "unnatural technology" would lead to dehumanizing systems. Ellul warned:

> The world that is being created by the accumulation of technical means is an artificial world and hence radically different from the natural world. It destroys, eliminates, or subordinates the natural world, and does not allow this world to restore itself or even to enter into a symbiotic relation with it. The two worlds obey different imperatives, different directives, and different laws which have nothing in common [7].

The 1950s were the very start of the Space Age and the birth of supersonic jets, an era when astrophysicists such as Edwin Hubble were helping humanity reach a deeper understanding of the size and dimensions of the cosmos. Thus, Ellul's ideas seemed strange and dark, and his warnings were largely ignored. Yet today, the idea of embracing green technology to save global biosystems, preventing global pollution, and preserving endangered species seems to be gaining ground with younger people.

Conclusions

This book provides a blueprint for the future. It explains how nature-friendly technology or environmentally compatible systems can provide new and better forms of employment for humans and foster new ways of thinking. Instead of technology that conquers nature, future technology would let humans live in a symbiotic relationship with nature. A number of notables, from Ralph Waldo Emerson to John Muir, sought to have technology connected with the natural cycles of life. Indeed, Muir famously said: "When one tugs at a single thing in nature, he finds it attached to the rest of the world" [8].

Technologist and inventor R. Buckminster Fuller found his greatest inspirations from observing and imitating nature. His famous geodesic dome, for instance, was based on observing naturally occurring crystalline structure. James Lovelock, a polymath scientist, environmentalist, and futurist who was 101 years old at the time of this writing in 2020, developed the idea of Earth as Gaia. He envisioned our planet as a sort of living, breathing, self-regulating community of organisms that interact with inorganic materials to sustain life. He argued that humans and human-developed technology should respect the community of animals and plants on Earth, since that synergy sustains all life on the planet. Despite Lovelock's concerns about technology polluting the world, he was ultimately optimistic that humans could see the errors of their ways and eventually evolve "Gaia-compatible" technology to sustain life on Earth [9] (Fig. 1.3).

Friend and colleague Eric Burgess, Science Editor for the *Christian Science Monitor* and now deceased, wrote the foreword to *Global Communications Satellite Policy*, this author's first book, back in 1974. He too was concerned with misuse of nuclear weapons and other technologies that might destroy humankind. He warned:

> Earth has an astronomical future as a habitable planet by probably 6 billion years. This is enough time, indeed, for a new race of thinking creatures to evolve once again from blue-green algae if all advanced terrestrial life forms become extinct. But the achievements of that a continuously evolving intelligent species might make in billions of years is unimaginable [10].

Technology might be needed to save humanity, but it may not be technologists who are best suited to defining what that technology should be and how it might be used. The best answers may well come from interdisciplinary and international teams. The ideas and solutions presented in this book rely in part on interviews and questionnaires posed to people in many fields from around world. There is great value in collective wisdom. As Lovelock once observed, modern humans often tends to mock the intelligence of the ancient Neanderthals, but he then posed this question: How many modern homo sapiens could successfully eke out a living in cave and survive off the land?

Fig. 1.3 Earth seen from the Moon. (Courtesy of NASA)

The political process must also be considered. What has the United Nations and the U.N. General Assembly done to address survivability issues? This entity, with technical advice and input from all over the world, first formulated what were called the "Millennium Goals." These goals were first set to be achieved in 15 years, or by 2015. After those 15 years, they drew on their experience and started again with a new plan. This set of comprehensive goals was reconceived in 2014 and 2015, and became the U.N.'s Seventeen Sustainable Development Goals for 2030. Approved by the world community, they are broad, interdisciplinary, and do generally touch on every aspect of life on the planet. These goals are also discussed later in this book. To date, they provide perhaps the most comprehensive and useful guide for the new types of survival technologies needed for a safer and more sustainable human future.

This introduction closes with the profound words of Astronaut William Anders, who said, "We came all this way to explore the moon, and the most important thing is that we discovered the Earth" [11].

References

1. Elliot, J.: A Football field-sized asteroid just missed Earth. No one saw it coming. Global News, June 15, 2020. https://globalnews.ca/news/7067086/asteroid-near -miss-earth/
2. The Climate Issue. Wired Magazine, April 2020
3. Friedman, M.: The social responsibility of business is to increase its profits. New York Times Magazine, September 13, 1970
4. Boccalandro, B.: Adam Smith: The Founder of CSR. https://www.beaboccalan-dro.com/adam-smith-founder-csr/. Last accessed 30 Sept 2020
5. Alex Gorsky, Chief Executive of Johnson & Johnson, as quoted in "The Thought Heard Round the World" New York Times Magazine, P. 2, September 13, 2020
6. NAE. NAE Grand Challenges for Engineering https://www.nae.edu/File. aspx?id=1. Last accessed 30 Sept 2020
7. Jacques Ellul Quotes. https://www.goodreads.com/work/quotes/266493-la-technique-ou-l-enjeu-du-si-cle#:~:text=%E2%80%9COur%20civilization%20 is%20first%20and,the%20situation%20is%20mere%20idealism.%E2%80%80% 9D. Last accessed 30 Sept 2020
8. John Muir quotes. https://www.google.com/search?sxsrf=ALeKk00qt9ZexH4zJ fEd-7jWNHYyIgUyQw%3A1601302879767&source=hp&ei=X_FxX9LWJce qytMPk8SU4AY&q=muir+quotes+about+nature&oq =muir+&gs_lcp=CgZwc3ktYWIQARgBMgUIABCRAjIECAAQQzILCC4 QsQMQxwEQrwEyBAgAEEMyBQgAELEDMgUILhCxAzIKCC 4QxwEQrwEQQzIECAAQQzIKCC4QxwEQrwEQQzICCAA6BAgjECc6 BQguEJECOggILhCxAxCDAToLCC4QsQMQxwEQowI6BwguELEDEEM 6BwgAELEDEEM6CAguEMcBEK8BOgQILhBDOgIILlD9DlimG2DhN mgAcAB4AIABpQGIAc8EkgEDMS40mAEAoAEBqgEHZ3dzLXdpeg&sclien t=psy-ab. Last accessed 30 Sept 2020
9. Lovelock, J.: The Vanishing Face of GAIA. Bassic Books, New York (2009)
10. Burgess, E.: Foreword to Joseph N. Pelton, Global Communications Satellite Policy, (1974) Lomond Systems, Mt. Airy
11. William Anders, Brainy Quotes. https://www.brainyquote.com/quotes/william_ anders_751620#:~:text=William%20Anders%20Quotes&text=We%20 came%20all%20this%20way%20to%20explore%2the%20moon%2C%20 and,that%20we%20discovered%20the%20Earth. Last accessed 30 Sept 2020

2

How Space Systems and New Technology Can Help

Humanity has only scratched the surface of its real potential.
　　　　　　　　　　　　　　　　　　　　–Peace Pilgrim

We cannot despair of humanity, since we ourselves are human beings.
　　　　　　　　　　　　　　　　　　　　–Albert Einstein

Introduction

It was contended in the preceding chapter that there are at least ten potential threats to the world and humanity. Yet, we are spending far more money and resources on growing the world economy than solving these problems. That the expansion of the economy is a higher priority than survival of the species is an idea that at least the young people of the world are beginning to question and find wanting.

If there is a lesson to be learned from the COVID-19 pandemic, it is that prevention of mega-disasters can not only help avoid major loss of life but also save trillions of dollars in economic losses. The cost of saving Earth is going up—probably exponentially so—and thus, many of the green programs that are being advanced make not just environmental sense, but economic sense as well.

Fortunately, there is still time to develop or expand existing technologies and systems. But conservation and recycling policies and regulations are needed. The conversion of the global economy into a sustainable mode of operation is also needed. The new technologies discussed in this book are not fail-safe approaches to the future.

Too often, the technology that would seem to rescue society today proves to be dangerous and polluting tomorrow. At the start of the twentieth century, the London City Council embraced the automobile as a marvelous means to rescue the streets of their fair city from tons of horse manure. In

J. N. Pelton, *Space Systems and Sustainability*, https://doi.org/10.1007/978-3-030-75735-9_2

London as of 1900, there were 11,000 hansom cabs and thousands of buses that required 12 horses each. A total of about 50,000 horses were producing 2 pints of urine and about 20 pounds of feces a day that were landing on London's streets. Automobiles were thus seen as a source not only of environmental rescue but also of economic relief due to the high cost of clearing the streets [1]. Ultimately, these benefits backfired as automobiles began polluting the city in other ways.

Unintended consequences must be examined before a technology is embraced on a wholesale basis. This is why vaccine trials must be extensive. This is why even new technology to further global sustainability must be carefully vetted. Yet despite these cautions, the time for change is now. Reform is urgent. Old and "dirty" technologies and industries must be retired. Clean systems with zero-carbon footprints must be implemented, with some due regard for transition away from badly outdated legacy systems. The cost of transitioning from coal-fired electrical plants may seem high, but the cost of recovery from greenhouse gas pollution is even higher. When we compute the cost of energy systems, we often forget to add in the cost of recovery and cleanup. Coal and nuclear energy's price tag is much higher than anyone first thought.

New space, information, energy, and environmental technologies, systems, laws, and regulations can help to reshape the world economy while restoring the environment and cleaning up the worst forms of pollution. Many of these cleanup operations are key to meeting the United Nations' Seventeen Sustainable Development Goals, which the General Assembly has set for 2030. The twenty-first-century "survival technologies" discussed in this chapter will be assessed against their ability to cope with these 17 prime objectives for improving the world in significant ways in this decade.

Care must be taken not to overstep. Too much of a supposed good thing can be a bad thing. New space systems for broadband communications and networking or for remote sensing and environmental monitoring would appear to be a very good thing. But over-deployment of such systems could lead to orbital space debris problems, loss of personal privacy, or major difficulties to scientific research such as electronic pollution and limitations on future radio and optical astronomy. Balance is key in almost all human enterprise.

These assessments draw on the wisdom of global experts from around the world. Advice and technical insight were sought from technologists, environmentalists, scientists, medical doctors, and experts from the fields of space, information, energy, defense systems, ecology, urban planning, and more. A concerted effort was undertaken to consult experts such as Lord Martin Rees (UK Royal Astronomer), Sir Martin Sweeting (CEO of Surrey Space

Technology Ltd.), Dr. Eric Viire (MD of the University of California San Diego and Arthur Clarke Center for the Human Imagination), Dr. Peter Martinez of South Africa (who heads the Secure World Foundation), Juan de Dalmau of Spain (President of the International Space University), Apollo Astronaut Rusty Schweickart, Dr. Donald Daniel (of the U.S. National Intelligence Council), and many other notables. These are experts not only on the latest technology but also on global threats, disarmament, climate change, pandemics, and more.

There are hundreds, if not thousands, of better pathways to the future that are called herein "survival technologies." These include new ways to detect pollution, forest fires, crop diseases, and forest infections. There are other eyes in the sky that can help detect illegal acts from prohibited fishing to smuggling and piracy. Other systems with new data analytics can help detect the onset and spread of pandemics and provide effective means to inform rural and remote peoples about how to access the best therapeutics against disease and infections.

Technology can be an instrument of peace, purification, and protection, or it can promote a disposable economy of thoughtless profiteering and pollution. As said many times already, it is not only a matter of developing key technologies but also of political action and regulation to prioritize sustainable systems of survivability.

The U.N. 17 Sustainable Development Goals for 2030

Global agreements are difficult in a time in which a number of so-called populist governments in democratic nations seem to be putting their own national interests first. Unwillingness to take in refugees or meet the goals of the Paris Accord on climate change, as well as sharp disagreements on trade agreements, punctuate global disharmony. Nevertheless, the General Assembly of the United Nations has drawn on leaders of the world to set important goals for global improvement by 2030. They are listed in Fig. 2.1.

The U.N. goals have indicated specific ways that progress toward them can be measured. Nine of the seventeen goals—more than half—are related in some way to coping with major existential risks as defined by this book. Thus, there are goals that address global pollution, cleaner energy, and more sustainable cities, threats to wildlife and plants, climate change, public health and improved global educational systems, and economic production and consumption.

Fig. 2.1 The U.N. 17 Sustainable Development Goals for 2030. (Source: Global Commons)

Goal 3 (health), Goal 4 (education), Goal 6 (clean water and sanitation), Goal 7 (affordable and clean energy), Goal 11 (sustainable cities and communities), Goal 12 (responsible consumption and production), Goal 13 (climate action), Goal 14 (life below water), and Goal 15 (life on land) are all very much on target.

Some of these goals, such as addressing hunger (Goal 1), decent work and economic growth (Goal 8), and industry, innovation and infrastructure (Goal 9), might in some circumstances work against global sustainability and the ability to create a cleaner and safer world. The key is to find a way to feed humanity and provide work and industrial expansion in a green and sustainable way. Such challenges might be for hypersonic transportation systems that are pollution-free, such as new types of mag-lev systems. The idea would be to create new forms of transport that are at least less dirty to the atmosphere than rockets and carbon-fueled jets. Perhaps, the most important and less costly would be widespread birth control, or perhaps less polluting crops like hydroponic farming.

The issue is how progress on most of the goals can be made compatible with constant economic growth and the need for jobs in 2200, when the global population may hit 12 billion. The issues of overpopulation and traditional, often "dirty" industrial expansion are perhaps the most difficult ones of the day. Nevertheless, new technology, just in the context of space applications, can make a significant impact. Annex 1 seeks to identify how space systems can help achieve these goals.

Another big omission from the U.N.'s list is how to address large natural disasters or cosmic threats. This was done perhaps in part because the scientific community itself has not yet embraced the idea that technologies might have reached the stage that they could now be initiated to address such threats successfully.

Now that we have identified the main threats and challenges facing humanity, the rest of this chapter provides a sneak peek at some of the technologies and systems that will be explored in greater depth throughout the book. This blueprint may have flaws and will need further refinements, but it is a good place to start.

Key Technologies to Prevent or Mitigate Outbreaks and Pandemics

The patterns of pandemics have changed dramatically over time. Centuries have normally passed between the outbreak of global pandemics over the past 2000 years. As the global population has expanded, the world has become more urban, and international travel has grown faster and faster, resulting in more and more outbreaks. There is a possibility that viruses have begun to mutate faster to defeat antibiotics. In the last century, since the Spanish Flu, there have been a record number of infectious disease outbreaks that either became pandemics or had the potential for massive amount of global deaths. Chapter 3 follows the history of pandemics and the development of modern medicine to understand how infectious diseases are transmitted and how their vaccines are developed.

Modern telecommunications, IT networks, and space systems can play a critical role in containing the spread of a new infectious disease. One method is to create one or more forms of a global network of hospitals and doctors that are networked together by communications satellites, fiber optic networks, and broadband cellular systems. Such health alert networks are designed to report quickly any new type of infectious disease that has suddenly appeared some place in the world. Thus, a global alert can be given. This same type of network is able to provide rapid updates on the progress of any threats to global health and the extent to which containment is being achieved.

There are new satellite constellations designed with very precise remote sensing systems. These networks are also connected in near-real time to high-speed data analytics. Such networks can help provide rapid containment of a new outbreak, even if a government decides not to report it or if doctors in a particular area are slow to understand that a new disease is spreading.

One such remote sensing satellite system has a large constellation of smart birds in the sky and is designed to obtain satellite imaging of the entire world every three seconds. This single system produces some 27 petabytes of data a day. The data analytics from this satellite network could be used to observe cars in parking lots to track customer patterns, or in the case of pandemics, it could track the number of cars and ambulances traveling to hospitals and clinics. An algorithm could sound an alert for any hospital or clinic that shows a rapid increase in traffic to that facility. The satellite network has very high resolution and unique capabilities that could allow it to spot changing patterns of behavior.

Another space capability would not halt the spread of the disease, but rather measure the economic impact. For instance, a new satellite constellation that measures radio frequency (RF) usage on a global basis, called Hawkeye 360, measured that shipping into Wuhan rapidly dropped to 40% of earlier levels during the outbreak of COVID-19 in that city [2].

Global Pollution

Tracking the location, type, and degree of global pollution is a very difficult task. With approximately eight billion people and hundreds of millions of industries and enterprises operating all over the world, in the oceans, the skies, and even remote arctic regions, their combined impact is not something that human inspectors or aircraft monitoring systems can do well on a global scale.

The latest remote sensing satellite systems with sophisticated data analytics can detect and measure the increase in ocean water acidity, pollution in lakes and rivers, the emissions of coal-fired power plants, and more. The most sensitive sensors can detect even small increases in pollution levels, and data analytics operating in near-real time can signal new sources of pollution much more quickly than any other type of monitoring system.

The effects of climate change, including parched, more arid regions and close to 50% more lightning strikes, are leading to much more frequent and extensive forest fires that are accelerating smoke pollution. Again, meteorological and remote sensing satellites, particularly those that can operate in near-real time, offer the most comprehensive alert system to detect remote forest fires or other incidents. Quick responses to oil spills or offshore drilling operation accidents can help contain the spread of oil slicks. It is not widely understood that oil contaminates eventually drift to the polar region and changes the albedo of the ice (darkens it), thereby accelerating ice thawing and leading to a more rapid increase in ocean water levels.

Monitoring global pollution activities faster, more accurately, and over the entire globe will help create a better warning system in the event action is needed immediately. Better incentives (i.e., new rewards for lessening the carbon footprint of power plants), enforcement of laws (i.e., fines and arrests), and improved technology for the reduction of greenhouse gases or the creation of carbon sinks are needed. These topics and more are evaluated in Chap. 4.

Biochemical and Nuclear Weapons

Biochemical and nuclear weapons could be considered a form of global pollution. Indeed, such weapons are the most dangerous type since they deliver toxic materials more quickly than conventional pollutants.

Today, there are satellite systems that are able to detect all types of nuclear explosions with great sensitivity. Some of these are hosted payloads on the Navstar GPS satellites, and others are on U.S. military satellites. There is currently no space-based system that is specifically designed to detect the release of nerve, mustard, or other poisonous gases or biological weapons. Nevertheless, there are remote sensing and surveillance systems that would be able to detect the effects of such a release.

There have now been cases where time-stamped images from remote sensing satellites have been used to prosecute crimes against humanity and human rights violations in the World Court of Justice and in the European Court on Human Rights. In some cases, the use of remote sensing images must first successfully show that individual rights to privacy have not been violated [3]. Other satellite capabilities can track legal and illegal use of radio frequencies to aid law enforcement, and also track drug smuggling and other offenses. Currently, the only commercial system to track radio frequency use globally is the Hawkeye 360 satellite constellation, deployed in 2019 and 2020 [4].

Shipments and ships in the world's oceans are now increasingly controlled by satellites that are able to track automatic identification system signatures for all vessels over 300 tons, as mandated by the International Maritime Organization (IMO). There are now nearly a dozen satellite networks that provide AIS services. This satellite monitoring capability has been invaluable for law enforcement and antiterrorist organizations. It would seem prudent for this monitoring to be extended, wherein AIS signatures are embedded in all nuclear weapons and containers associated with biochemical weapons as part of any future arms control treaty [5]. Methods for control and inspection techniques of biochemical and nuclear weapons are examined in Chap. 5.

Potentially Hazardous Asteroids

Coping with potentially hazardous asteroids (PHAs) has at least two dimensions. One aspect is designing space and ground observatories capable of detecting asteroids that might hit Earth. The other is agreeing on the minimum size of asteroids that are included in the search profile, as we saw in Chap. 1.

The problem is complicated. A survey to locate all 35-m versus 140-m PHAs represents a big difference in the effort involved. Asteroids that are 35 m in diameter are not four times smaller than those that are 140 m; these so-called city killers are actually 64 times smaller. Further, they are somewhere between ten to a hundred times more numerous. The difficulty of spotting smaller but still deadly asteroids is quite demanding, but there are increasingly clever means to do so, which suggest that lower cost, smaller infrared space telescopes might still be able to do the trick.

The true dilemma is finding a reliable and economically viable way to divert a PHA, so it misses Earth when it is actually found. This is something we must get right since a small diversion might end up hitting Earth in a more vulnerable location. If a threatening space rock is located early with years of warning, there might be many low-cost, reliable solutions. On the other hand, if the threat is only detected at the last minute, the options are decidedly less.

New ideas are most welcome. Fortunately, organizations such as the B612 Foundation, the Planetary Society, NASA, and many other space agencies, as well as the COPUOS, the IAWN, and the SMPAG, are all aware of the challenges and trying to find solutions. These are explored in greater detail in Chap. 6.

Comets and Other Cosmic Threats

One of the most frightening cosmic scenes that ever transpired took place in July 1994, when over 20 pieces of the Shoemaker-Levy comet slammed into the surface of Jupiter. The world's scientists were watching as various parts of this comet, labeled A through W, smashed into the giant planet between July 15 and July 22. This event was captured in detail via the Hubble Space

Telescope, the Galileo orbiter, the Ulysses, the Voyager 2 space probe, and Earth-based telescopes. The force of these fragments hitting Jupiter was calculated to be equivalent to the force of 300 million atomic bombs and created huge dust plumes that rose as high as 3000 km above Jupiter's surface. One of these impacts alone would have represented an existential extinction event for humanity if they had hit Earth instead. The dust plumes that blanketed Jupiter were thought to be similar to those that circled Earth after the huge asteroid hit our planet 65 million years ago and wiped out the dinosaurs [6].

Fortunately, Earth has two major defensive systems against comets. These are the huge gravitational fields of the Sun and Jupiter, which are most likely to capture comets instead. Indeed, the Shoemaker-Levy comet was thought to have been captured by Jupiter's gravity for 10 years before various parts finally descended to the surface. This is very fortunate because the size and velocity of comets make them much harder to defend against than asteroids.

There are other significant technical challenges from cosmic threats where serious study is needed. The biggest is actually caused by humans. The careless launch of many thousands of satellite missions and the abandonment of the upper rocket stages in orbit have now given rise to a growing problem of orbital space debris in low Earth orbit, medium Earth orbit, and geosynchronous Earth orbit. There is now over 11 metric tons of space debris in orbit, and collisions of major space objects are creating an ever larger number [7]. Plans to launch tens of thousands of satellites into large constellations, especially in low Earth orbit between 700 km and 1200 km, is a mounting concern. There is serious concern as well about the need for either Active Debris Removal (ADR) or "recycling" of debris elements into useful space objects.

These issues and the technical challenges associated with them are addressed in Chap. 7 [8].

Solar Storms and Coronal Mass Ejections (CMEs)

As we have seen in Chap. 1, the occasional eruptions from the sun, known as coronal mass ejections, can send a massive number of ions out from the sun's surface, traveling at millions of kilometers an hour. If the eruption occurs in just the right way, these ions will form a massive pathway of destruction, smashing through the Earth's natural magnetic shielding. The result would knock out electrical power grids, burn out transformers, destroy the controls on pipelines, put satellites out of commission, and possibly jeopardize the world's telecommunications and computer networks. In a world where virtually everyone relies on electricity, the world's economy would come to a standstill. People would not be directly harmed, but the world's entire economy, its defense systems, its banks, its lines of supply, and its airlines could suddenly go caput.

Much more research is needed to protect against such an event. Some protection can be undertaken using heavy-duty circuit breakers, powering down satellites, and putting vital network controls inside of so-called Faraday cages. More exotic ideas are the creation of solar shields at Lagrange Point 1. The technology is likely now available to allow a magnetic shield to be built for Earth. Such a shield might require billions of dollars to be constructed, but if done right, it could save trillions of dollars. A magnetic shield could also be built for Mars. This could stop the solar wind from stripping away Mars' natural environment and allow an atmosphere to evolve. Megastorms from the sun come about only once every 150 years. This interval, although quite long, is much shorter than the span of time between civilization-ending asteroids or comets.

Then, as we have seen, explosion detection systems that are in place today onboard spacecraft would have a difficult time discerning the difference between a solar-triggered EMP and a nuclear blast-triggered EMP. Both of these, one natural and one manmade, would be detected at high altitude. The killing effect on electronics in the area of impact would be very much the same. The questions thus become: How does a detection device distinguish between the sources? What is the range of possible responses to restore electrical power systems? How can defense systems be sufficiently protected against both types of events?

All of these concerns are considered in Chap. 8.

Population Expansion

The exponential population growth as shown in Fig. 2.2 has occurred most significantly in the last 300 years. This rise has been made possible due to agricultural development and the Industrial Revolution. The pattern of zero population rates in many economically developed countries is reassuring to many demographic experts. China and Singapore seem to be the only countries that have managed to impose population controls with any degree of success; in both cases, the result has been significant income increase per capita.

Table 2.1 shows this growth trend into the future. The result would truly catastrophic. There would be environmental devastation, economic chaos with massive levels of unemployment, ruinous exploitation of resources and mass extinction of wildlife, loss of much of the world's oxygen supply, death

World Population Growth Through History

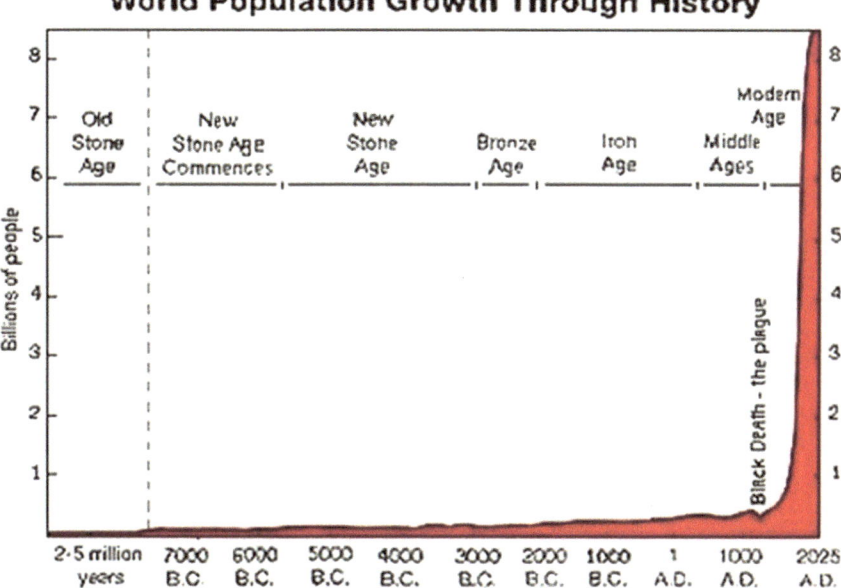

Fig. 2.2   The rapid rise of human populations

Table 2.1   Future growth of human population

Year	Global population
1800	0.8 billion
1900	1.8 billion
2000	7.0 billion
2100	12 billion??
2200	24 billion??
2300	48 billion??

to life in the oceans, and unbelievable problems with housing, education, health care, transportation, power supply, urbanization, and water supply. Overall not a pretty picture. Overpopulation is perhaps the root cause of the other existential threats to the world.

Perhaps, the key technologies of the twenty-first century may come in such fields as birth control, family planning, or urban planning, to allow more people to live in a sustainable way at higher levels of density. Other key technologies might include better ways to globally monitor and control expansion of human settlements into wildlife areas, the oceans, and other locales vital to sustaining life on Earth. The very difficult and sensitive issue of global population control and related innovations needed to achieve progress in this area are addressed in Chap. 9.

Climate Change

The range of issues presented by climate change are so large, significant, and diverse that one almost does not know where to begin. The range of necessary technologies, monitoring systems, enforcement, and regulatory systems is also very difficult to comprehend. Most of the power and capability to address these issues reside at the governmental and corporate level.

Aspirational goals such as those in the Paris Agreement are not sufficient. Yes, explicit worldwide goals are needed, yet these need to be backed up by national commitments reflected in national laws with specific targets, rewards, and penalties. In no other area is the need for institutional and regulatory reforms so overwhelming large. It is strongly felt that there is a need for a new mechanism such as a Global Sustainability Treaty, as well as new commitments by the top leaders of the world, especially within the G20 group. This is key in that any consensus within the G20 would represent agreement among the countries responsible for over 90% of the world's energy consumption, pollution, and global population. If there can be consensus within the leaders of these countries on climate change objectives, then true headway can be achieved.

Chapter 10 thus addresses climate change with a particular focus on environmental, space systems, and data analytics. Only a space-based environmental policing system in the sky can provide the oversight essential to comprehensive enforcement—or in some cases rewarding those who meet their ecological objectives.

Natural Disasters

For millennia, humans have viewed natural disasters as inevitable events. Such natural catastrophes were to be prepared for, responded to by rescue efforts and rebuilding activities, and then preparations for the next disaster began anew. This cycle would never stop, as the natural makeup of Spaceship Earth included a wide range of inherently destructive forces, such as earthquakes and tsunamis, volcanoes and super volcanoes, hurricanes, typhoons, cyclones, tornadoes, tropical storms, flooding, forest fires, and avalanches. Of course, there were ways to mitigate such disasters. Architects became more sophisticated in building structures that might withstand most earthquakes. Dams, levees, and tidal basins were built to control flooding. Many structures could be built to withstand violent storms. Lightning rods and fireproofing could provide some protection against fires.

The question raised in this book is whether human technology has reached a level where more sophisticated technology could mitigate natural disasters to a much larger degree. Some of these actions do not require great sophistication. Well-developed technology at the Pacific Disaster Center in Hawaii can model flooding patterns that can occur due to strong tidal waves and tsunamis, showing lands that should be left as parks and natural areas that should not be built on in Hawaii and many shore areas of the Pacific. In short, contemporary computer modeling can clearly show the places that represent danger areas due to potential flooding, earthquakes, and volcanic activity. The systematic use of such modeling to identify extreme risk natural disaster areas and subsequent governmental action to compensate and perhaps relocate inhabitants should proceed.

In addition to this, Chap. 11 explores technologies such as solar shields, new types of space systems that could help moderate extreme weather systems, and systems that might prevent or slow the melting of the ice or frozen peat fields in arctic regions.

Artificial General Intelligence and Cyberattacks

Astrophysicist Stephen Hawking, SpaceX's Elon Musk, Google's Jack Dorsey, and other notables issued a letter in 2016 warning about the danger of developing AI "autonomous weapons" systems that might make independent life-and-death decision. Hawking stated the concern in stark terms: "The development of full artificial intelligence could spell the end of the human race." [9]

These concerns are wide ranging and complex. They include such concerns as massive cyberattacks against vital infrastructure, underemployment due to super automation, and actual conflict between AI and human intelligence. The dangers from the most sophisticated forms of self-aware artificial intelligence are indeed profound. These are examined in some depth in Chap. 12.

The Global Sustainability Treaty

This examination considers not only environmental, cosmic, and other dangers but also the needed reforms in technology, goals, practices, global alliances, and economic, governmental, legal, and regulatory policies. A "Global Sustainability Treaty" could allow global cooperative actions in all of these areas. Such a treaty would not create a new international agency. Rather, the objective would be to knit together all of the global, regional, and national agencies that work on various aspects of these existential threats, creating interdisciplinary databases and fostering a global approach. A first step would be to create a global task force drawn from existing institutions, which would define a new agenda for change and address key sustainability and survival objectives. All this is outlined in Chap. 13.

Conclusions

The first step toward global sustainability and long-term survival of humanity is to recognize the breadth and depth of the various challenges to human civilization as we move into the 2020s and 2030s. There are over 30 new technologies and systems outlined in this book that can help us do so. Annex 1 notes all those that could assist with the United Nations' 17 Sustainable Development Goals for 2030. These are just some of the innovations that could help, but they provide a useful summary of key space innovations.

Annex 1: Assessment of How New and Existing Space Application Can Achieve Sustainable Development Goals

UN sustainable development goal	Telecom and networking Sats plus high-altitude platform systems (HAPS)	Broadcasting Sats and high-altitude platform systems (HAPS)	Remote sensing and Earth observation Sats	Meteorological Sats	Navigation & timing Sat	Solar power Sats
Zero Poverty (Goal 1)	New jobs via telework, opportunity for remote services, tele-training in remote villages	Broad distribution of information on birth control, nutrition, vaccines, etc.	Improved information to support fishing, farming, forestry, mining, etc.	Reduced loss of crops and trees; better response to climate change	Improved farming and fishing via precision geolocation	New capability in the future
Zero Hunger (Goal 2)	More efficient agricultural & fishing processes	Broad distribution of information on nutrition & birth control	More productive farming & lower cost food	Less crop loss due to unpredicted storms, flooding, typhoon, hurricanes	Improved farming and fishing via precision geo-location	New capability in the future
Good Health and Well-Being (Goal 3)	Tele-health and remote medical service	Broad distribution of information on birth control, nutrition, vaccines, etc.	Detection of crop or tree disease; ability to detect pandemics	Detection of solar flares & ozone holes that cause skin cancer & gene damage	Ability to precisely track spread of disease & pandemics	New capability in the future

(continued)

(continued)

UN sustainable development goal	Telecom and networking Sats plus high-altitude platform systems (HAPS)	Broadcasting Sats and high-altitude platform systems (HAPS)	Remote sensing and Earth observation Sats	Meteorological Sats	Navigation & timing Sat	Solar power Sats
Quality Education (Goal 4)	Quality tele-education programs, remote testing, & degree programs	Educational radio & television; access to global news	Less destruction of schools & educational infrastructure	Less destruction of schools & educational infrastructure	Cost savings on school transportation	New capability in the future
Gender Equality (Goal 5)	Tele-educational programming & access to databases	Access to global news and TV broadcasts	Access to more types of jobs	To be determined	To be determined	To be determined
Clean Water (Goal 6)	Tele-education on water purification & sanitation, tele-training of engineers and scientists	Broadcasts on water purification & sanitation	Detection of polluted water & acid rain; road access information for water trucks following disasters	Better protection of water reservoirs against storms	Locate polluted waters; tracking of storms, tracking of acid rain, etc.	To be determined

Affordable and Clean Energy (Goal 7)	Tele-education, Internet access to more info on solar, wind, tidal, geothermal, OTEC, & other clean energy systems	Broadcasts on energy savings & building of clean energy systems	Aid in finding good locations for wind farms, geothermal energy, & tidal energy systems	Aid in finding good locations for wind farms, geothermal energy, & tidal energy	Assist in location of renewable energy systems	In future: Lower cost clean energy to cities, rural, and remote locations
Decent Work and Economic Growth (Goal 8)	Telework, village-based training, tele-banking, tele-services	Open university training	Aid to more productive, greener mining, fishing, farming, forestry, & transport	Support new construction & design of infra-structure related to climate change	Support for new construction & design of transport systems	In future: Lower cost clean energy to cities, rural and remote locations
Industry, Innovation and Infrastructure (Goal 9)	Tele-education, Internet-based innovation, Internet-based technology incubators, protective security for infrastructure	Educational radio & television; access to global news and specific programs on innovation	Aid to more productive mining, fishing, farming, forestry, and transport	Support to new construction & design of infra-structure related to climate change	Support to new construction & design of transportation systems	In future: Lower cost clean energy to cities, rural, and remote locations

(continued)

(continued)

UN sustainable development goal	Telecom and networking Sats plus high-altitude platform systems (HAPS)	Broadcasting Sats and high-altitude platform systems (HAPS)	Remote sensing and Earth observation Sats	Meteorological Sats	Navigation & timing Sat	Solar power Sats
Reduced Inequalities (Goal 10)	Tele-education, Internet-based learning & access to databases	Educational radio & television, access to global news	To be determined	To be determined	To be determined	To be determined
Sustainable Cities and Communities (Goal 11)	Substitution of tele-services and tele-work for physical transportation to work; access to broadband services	Educational radio & television; access to global news	Key topographic information for transportation, water, and sewer planning	Key information related to protection of city infrastructure from violent storms	Improved traffic & transportation control	In future: Lower cost clean energy to cities, rural, and remote locations
Responsible Consumption and Production (Goal 12)	Tele-education, tele-work & other satellite services can be provided worldwide & especially in rural areas	Broadcasts on tele-work, conservation, energy savings, & building clean energy systems	Monitor hazardous waste locations, atmospheric & ocean pollution, oil spills, garbage scows, etc.	Note changes in weather & climate due to industrial activities	Accurately pinpoint sources of pollution	In future: Lower cost clean energy to cities, rural, and remote locations

Climate Action (Goal 13)	Tele-education, tele-work, tele-services, clean satellite services can be provided worldwide	Broadcasts on tele-work, conservation, energy savings, & building clean energy systems	Track ice cap & glacier melting; measure ocean and atmospheric temperatures & acidity	Track changes in atmospheric temperatures, intensity of storms, solar activity	Pinpoint location of atmospheric & oceanic sensors	In future: Lower cost clean energy to cities, rural, and remote locations
Life below Water (Goal 14)	Tele-education, global Internet access; track location of endangered species	Satellite broadcast TV & radio can strengthen education, civic activism, and knowledge of law	Detection of water & ocean pollution & acidity, coral bleaching, fish, algae, & plankton depletion, etc.	Track ocean storms & hurricanes	Determine exact location of sensor & ocean buoys	To be determined
Life on Land (Goal 15)	Tele-education, global Internet access; track location of endangered species	Satellite broadcast TV & radio can strengthen education, civic activism, & knowledge of environmental law	Track animals & endangered species	Monitor violent storms and provide flood & high wind warnings	Determine exact location information of earthquakes, volcanos, and coordinate rescue operations	To be determined
Peace, Justice and Strong Institutions (Goal 16)	Low-cost satellite telecommunication & Internet access can strengthen education, civic activism, and knowledge of law	Satellite broadcast TV & radio can strengthen education, civic activism, and knowledge of law	Time-stamped remote sensing data has been used to prosecute crimes against humanity	To be determined	To be determined	To be determined

(continued)

(continued)

UN sustainable development goal	Telecom and networking Sats plus high-altitude platform systems (HAPS)	Broadcasting Sats and high-altitude platform systems (HAPS)	Remote sensing and Earth observation Sats	Meteorological Sats	Navigation & timing Sat	Solar power Sats
Partnerships for the Goals (Goal 17)	Satellite manufacturers & service providers can help promote telework, tele-education, and tele-health	Satellite manufacturers & satellite broadcasters can help promote tele-work, tele-training, tele-health, & tele-services	To be determined	To be determined	To be determined	To be determined

Note: This table is copyrighted by Joseph N. Pelton and licensed to Springer Press for this publication.

References

1. Ben, J.: The Great Horse Manure Crisis of 1894. https://www.historic-uk.com/HistoryUK/HistoryofBritain/Great-Horse-Manure-Crisis-of-1894/. Last accessed 21 Oct 2020
2. Case Study. Hawkeye 360 Reveals Chinese Vessel Activity Plummeted during Covid-19 Outbreak. https://www.he360.com/hawkeye-360-reveals-chinese-vessel-activity-plummeted-following-covid-19-outbreak/ (2020)
3. Ana Cristina Núñez M.: Admissibility of remote sensing evidence before international and regional tribunals. https://www.amnestyusa.org/pdfs/RemoteSensingAsEvidencePaper.pdf (2012)
4. About Hawkeye 360. https://www.he360.com/about/. Last accessed 10 Oct 2020
5. European Space Agency. Satellite – Automatic Identification System (SAT-AIS) Overview. https://artes.esa.int/sat-ais/overview. Last accessed 20 Oct 2020
6. NASA, Solar System Exploration. P/Shoemaker-Levy 9. https://solarsystem.nasa.gov/asteroids-comets-and-meteors/comets/p-shoemaker-levy-9/in-depth/. Last accessed 21 Oct 2020
7. Pelton, J.N.: New Solutions for Orbital Debris Removal. Springer Press, Switzerland (2016)
8. Pelton, J.N.: Orbital Space Debris and Other Threats from Outer Space. Springer Press, Switzerland (2013)
9. Cellan-Jones R.: Stephen Hawkins warns artificial intelligence could end mankind. BBC News, December 2, 2014, https://www.bbc.com/news/technology/30290540

3

Pandemics

The aim of medicine is to prevent disease and prolong life; the ideal of medicine is to eliminate the need of a physician.
    –William J. Mayo, M.D. of the Mayo Clinic

True prevention is not waiting for bad things to happen; it's preventing things from happening in the first place.
    –Don MacPherson

Introduction

The world of space is changing faster than ever. The satellites now being launched have reached dizzying heights. It is not the heights themselves that are dizzying, but rather the number of satellites involved. These new types of satellites are being deployed into low Earth orbit in large-scale constellations, often numbering in the thousands. They are used primarily for communications and remote sensing purposes. There are, for instance, two new satellite systems for broadband communications being launched by SpaceX. When and if they are fully deployed, these two networks alone might total over 10,000 satellites.

The high-resolution remote sensing satellite constellations of the near future will provide coverage of the entire Earth. Supported by amazingly fast data analytics and artificial heuristics, these will soon be able to provide precise images of the Earth with over a quadrillion data points in one collective global image with near-instantaneous updating. While more satellite communications can bring more broadband access, global education, and service jobs to the world, more remote sensing can help humans better manage the health of the planet and humanity as well.

This unprecedented level of global remote sensing data can prove vital to tracing population growth and detecting disease outbreaks. The most advanced satellites may be critical to studying disturbed habitats that can give rise to *zoonotic diseases*, animal-based, infectious diseases that spread to humans, where often the human physiology has no existing antibody defense. It is this medical phenomenon that can give rise to an alarming number of global pandemics.

J. N. Pelton, *Space Systems and Sustainability*, https://doi.org/10.1007/978-3-030-75735-9_3

Perhaps, the most important capabilities will not come from the satellites themselves but from the data analytics and AI software that can interpret the remote sensing data faster and transfer the results more quickly to other parts of the world through improved global connectivity. The latest in these satellite digital networking and communications systems are LEO constellations that are designed to bring low-cost, broadband connectivity to the underserved areas of the world. Again, new networking architectures and software can allow more rapid transfer of the latest information about infectious disease outbreaks and their rate of spread. These can also be important for transferring information on the latest therapeutics or updates on the availability of vaccines. Other applications might include information regarding the efficacy of therapeutics or vaccines, or the closest location where necessary medical assistance can be provided.

Viruses and Vaccines

Coronaviruses are respiratory diseases and include the common cold. There are actually four different types of coronavirus groups. These are divided into alpha, beta, gamma, and delta. The following coronaviruses can infect humans: 229E (alpha); NL63 (alpha); OC43 (beta); HKU1 (beta); MERS-CoV, a beta virus that causes Middle East respiratory syndrome (MERS); SARS-CoV, a beta virus that causes severe acute respiratory syndrome (SARS); and SARS-CoV-2, which causes COVID-19. The last three of these viruses have proven to be very dangerous. All three are suspected zoonotic diseases that jumped from an animal to a human.

This jump is not restricted to MERS, SARS, and COVID-19. The Ebola virus made the jump as well. This is not a respiratory virus spread via coughing or sneezing, but rather a virus that spreads through fluids. It took decades, but finally, cooperative work across three continents has led to a successful Ebola vaccine.

One of the first vaccines developed to combat the COVID-19 virus originated with Pfizer pharmaceuticals. Similar viruses are now being fully tested and delivered to hospitals from vaccine developers Moderna, Johnson and Johnson, AstraZeneca, and others. The Pfizer vaccine was the first, but certainly not the last. The Pfizer vaccine is designed to instruct the human body on how to create antibodies to fight the COVID-19 virus. These instructions are based on a specially designed messenger RNA (mRNA) cell.

This new wave of vaccines uses tailor-made genomes designed through Crispr gene-editing technology. The tactic is proving to be almost as groundbreaking as the first vaccines developed by Dr. Louis Pasteur over a century ago. In theory, the Crispr gene-editing process and the mRNA delivery system approach can be used again and again to develop future vaccines against other zoonotic viruses. Unfortunately, this is a future that now seems almost guaranteed. Indeed, the COVID-19 virus has already begun mutating, with new strains that seem to allow even more rapid spread [1]. Just before the Christmas holidays in 2020, mutated strains had been detected in the United Kingdom, the Netherlands, Australia, and Malaysia. As of January 2021, these have spread throughout the world. The one thing that is clear is that the COVID-19 zoonotic virus is not stable, and this will perhaps be characteristic of other zoonotic viruses in years to come.

The hope for the future is that detailed satellite monitoring systems can do one of two things. The first would be to detect zoonotic diseases more rapidly. The hope is that with sufficiently early detection, it might be possible to contain and seal off their spread. Secondly, these systems may generate quicker understanding of the nature of the latest outbreak, allowing more rapid Crispr-designed, RNA-based vaccine development. The exceptional data analytics and heuristic software available on these satellite systems are revolutionary. The latest data transfer capabilities and super-computer processing speeds can be used to find telltale signs of diseases concealed within high-resolution satellite data.

Experience will show which algorithms are the most effective in combatting disease outbreaks. Heuristic analytics may someday soon be able to detect a surge of people suddenly going to hospitals, clinics, and doctor's offices. Perhaps, new triggers can be developed through ground-based sensors. The key will be to create new, better, and faster systemic alerts. Certainly, this is helped by the spread of low-cost, highly mobile telecommunications user devices such as 5G phones and Internet of Things (IoT)-equipped devices that can eventually blanket the world. We must keep in mind, however, that new space capabilities can also make some problems worse if used to increase activities or trends that could quicken the spread of disease—for instance, if space tools indirectly bolster population growth or further disturb native habitats. With technology, there is often one step forward, followed by one step back.

This chapter starts by tracing some of the important medical discoveries throughout history, to demonstrate how resistant humans are to changes in established practices and belief systems. Ultimately, only courageous research and willingness to carefully use experimental scientific practices can move medical practices forward. Then, the chapter examines the global response to various coronal viruses in recent decades. It explores the rise of zoonotic diseases and the global need for much better understanding of modern pathogens. The path forward must be more global cooperation to extend what might be called medical sustainability efforts across the planet. This entails new tracking technology using the Internet, smartphone capabilities, and telecom and remote satellite systems.

A Global Approach

Our society is ever more urban and interconnected. It is held together more closely by aircraft and international trade. Dr. Brian Bird, a research virologist from the University of California, Davis, who has studied the spread of Ebola in rural Sierra Leone, stated the problem in this manner: "The risks are greater now. They were always present and have been there for generations. Diseases are now more likely to travel further and faster than before" [2]. Yet, an Intelsat initiative organized by the author in 1987, called Project Share, shows the promise of what this same interconnectedness might bring. SHARE stands for Satellites for Health and Rural Education. The global video-seminar included some 67,000 doctors, researchers, nurses, and health caregivers in North and South America, plus Europe and Africa. The world's top experts on the AIDS/HIV coronal virus were connected together via an elaborate multisatellite videoconference hookup. This activity was initiated by the Miami Children's Hospital and organized over several months. At the event, attendees shared the latest information on AIDS/HIV research. That meeting was over 30 years ago. Nowadays, this type of project could be accomplished on a much more global scale in the new age of satellite constellations and network interconnectivity. A global MEDNET satellite system and database for medical treatment, research, and pandemic alerts could revolutionize the world of medicine—for research, for updates on therapeutics and vaccines, and for virus alerts, among dozens of other applications. A global network for education with many different channels and in many different languages was one of many aspirations of this author that have not been fully realized—at least in his lifetime. There is always hope.

Historical Background

The history of disease and infections is as old as the recorded history of humanity. Yet even today, in a world in which the human genome has been decoded and modern medical and scientific discoveries made, knowledge about infections has sometimes been slow. Amazingly, many people do not accept the most basic protocols to avoid the spread of this disease. Misinformation about the importance of self-isolation, wearing masks, and effective social distancing has been an ongoing problem with the current COVID-19 virus. Widespread vaccine inoculation remains a challenge, as there remain theories about how vaccines might give rise to autism or other chronic diseases, although there is no medical confirmation of such beliefs. All of these problems remain barriers to the treatment of illnesses and the fight against disease and pandemics. Good communications and a clear-cut tracking system can help contain a pandemic and get medicines and vaccines where they are needed quickly and more efficiently.

The advent of zoonotic diseases and global transportation systems are concerns unique to the twentieth and twenty-first centuries. The medical research community has found that pathogens and viruses seem to have evolved with humanity, becoming more adept at leaping from animals to humans as well as becoming resistant to antibiotic drugs. Pathogens are explored in greater detail in Chap. 11.

Once these pathogens emerge, they also seem to mutate more quickly. In short, the modern world has seen the evolution of pathogens that are "smarter," more aggressive, and more easily spread among an ever-larger global population. The more people that populate Earth, the more vulnerable human civilization becomes. This is a correlation that will be explored in greater depth in Chap. 9.

It is useful to review a very condensed history of modern medicine and medical treatment right up to today's efforts to map the human genome, explore the use of T cells, and develop modern vaccines. This section is meant to illustrate why there is such a pressing need for more global cooperation through medical research and treatment. This covers not only therapeutics and vaccines but also the larger issue of modern pandemics and sustainability. Such cooperation would go beyond the activities of the World Health Organization (WHO).

Medical and scientific research and containment methods have been used to contain deadly pandemics that might have caused millions of deaths but did not do so. Table 3.1 depicts all pandemic outbreaks over time since 165.

Without containment, these numbers would have been much higher. It is of course hoped that COVID-19, although not contained, will eventually be controlled and deaths held to the range of one to two million. This is the good news. The bad news is that the frequency of pandemic outbreaks has risen precipitously. Since the Spanish Flu, there have been eight potential global pandemic outbreaks. This is a record by far. Over the last 18 centuries,

Table 3.1 Global pandemics and containment

Name of pandemic	Time of occurrence	Type/prehuman host	Number of deaths
Antonine Plague	165–180	Believed to be either smallpox or measles	5 M
Japanese smallpox epidemic	735–737	Variola major virus	1 M
Plague of Justinian	541–542	*Yersinia pestis* bacteria/rats, fleas	30–50 M
Black Death	1347–1351	*Yersinia pestis* bacteria/rats, fleas	200 M
New World Smallpox Outbreak	1520 onwards	Variola major virus	56 M
Great Plague of London	1665	*Yersinia pestis* bacteria/rats, fleas	100,000
Italian Plague	1629–1631	*Yersinia pestis* bacteria/rats, fleas	1 M
Cholera Pandemics 1–6	1817–1923	*V. cholerae* bacteria	1 M+
Third Plague	1885	*Yersinia pestis* bacteria/rats, fleas	12 M (China and India)
Yellow Fever	Late 1800s	Virus/mosquitoes	100,000–150,000 (USA)
Russian Flu	1889–1890	Believed to be H2N2 (avian origin)	1 M
Spanish Flu	1918–1919	H1N1 virus/pigs	40–50 M
Asian Flu	1957–1958	H2N2 virus	1.1 M
Hong Kong Flu	1968–1970	H3N2 virus	1 M
HIV/AIDS	1981–Present	Virus/chimpanzees	25–35 M
Swine Flu	2009–2010	H1N1 virus/pigs	200,000
SARS	2002–2003	Coronavirus/bats, civets	770
Ebola	2014–2016	Ebolavirus/wild animals	11,000
MERS	2015–Present	Coronavirus/bats, camels	850
COVID-19	2019–Present	Coronavirus—unknown (possibly pangolins)	776 K (Johns Hopkins University estimate as of 8:27 am PT, Aug 17, 2020)

Source: https://www.visualcapitalist.com/history-of-pandemics-deadliest/

pandemics have been rare, a century apart or more. But this is no longer typical in today's world. We now live in a society populated by many billions of urbanized people interconnected via jet, train, and boat. Modern pandemics have the potential of not only bringing a staggering number of deaths but also huge disruptions to the global economy [3].

Germ Theory from Leeuwenhoek to Pasteur to Lister

Perhaps, the greatest discovery in medicine was that of microscopic organisms and bacteria. The Dutch scientist Leeuwenhoek is considered the father of microscopy for his study of microscopic organisms and bacteria in the seventeenth century. In a letter to the Royal Society in 1676, he informed them of his detection, wherein he had observed specialized cells that exhibited all the characteristics of life. Scientific inquiry into these microorganisms was subsequently pursued by a number of people, including Linnaeus, who published *System Naturas* in 1735, as well as Richard Bradley, Benjamin Marten, and Jean-Baptiste Goiffon.

It was Dr. Louis Pasteur (Fig. 3.1), who lived from 1822 to 1895, who unlocked the way forward for modern medicine. Most of all, Pasteur realized that the ancient belief that disease was the result of "spontaneous generation" was indeed quite wrong. He understood that infected wounds were caused by bacteria and realized the importance of sterilization for successful surgeries. His understanding led to the cornerstone practice of sterilization, further comprehension concerning the nature of pandemics and vaccine, and even the pasteurization process to remove bacteria from foods.

Early on, Dr. Pasteur studied the fermentation process in wine and beer. He had been hired by industrial producers of beer to understand little understood "beer disorders." Through this research, he was able to sort out the "living matter" associated with yeast production from inorganic crystalline structures. He recognized the chemical difference between living and dead matter,

Fig. 3.1   Dr. Louis Pasteur, known as the father of modern medicine and as the "Angel of Science". (Source: Wikimedia Commons)

between organic hydrocarbon chains that contained life at the microscopic level and inorganic crystalline structures that didn't. He studied the difference in the chemistry and makeup of microbes used in the fermentation process from the bacteria seen in the putrefaction or rotting process. In a series of experiments, he made observations about an epidemic caused by bacteria common to silkworms and determined how specific bacteria were the root cause of the infection process. This led him to find a vaccine that could be used to protect chickens from fowl cholera. And so Pasteur invented the process of immunology. This in turn led him to create workable vaccinations that could be used to prevent both rabies and anthrax [4] (Fig. 3.1).

Cholera, Dr. John Snow, and Dr. William Farr

Many of us are skeptical of what they cannot see, hear, smell, or touch. This is why things like electricity, magnetism, gravity, bacteria, viruses, or atomic structures are hard for people to fully comprehend. When out of the blue, hundreds or even thousands of people started mysteriously dying without any visible cause, there was great unrest among societies. Many people started to look to the supernatural. When babies are born, they are said to be given the "breath of life." Thus, the idea that disease might come from the "bad breath" of a sickness was prevalent for a very long time. It took an even longer time to learn that some bacteria could indeed be transmitted either through the air, as in a sneeze or cough, or through water or fluids.

When so-called Asian Cholera spread to Europe in 1831, the prevailing belief was that the spread of the disease was through the air. A subsequent cholera epidemic outbreak occurred in London in 1854. Dr. John Snow, a physician, carefully plotted an incidence chart of where the most cases occurred. He found they were clustered in Soho, an area heavily populated with cow sheds, slaughterhouses, and the droppings of horses. The statistical results showed that the greatest amount of sickness was clustered around the Broad Street water well in Soho. Dr. Snow presented his findings to the City Council and carefully explained his "unorthodox new hypothesis" that fecal materials had infected the pump and pump handle. He also confirmed that people who were supplied from the two water companies that directly obtained their water from the polluted Thames were most susceptible to getting cholera.

The City officials rejected Dr. Snow's findings. So, he committed an act of civil disobedience, removing the Broad Street pump handle by himself. Thereafter, the Cholera epidemic subsided. The London City Council and Mayor then reinstalled the Broad Street pump handle and officially rejected the "unpleasant idea" that fecal material at the pump could have spread the disease. This raging sickness is now known as the Broad Street or Golden Square Cholera epidemic.

At the time, Dr. William Farr agreed that the disease was clustered in Soho. Yet, he continued to hold the more popular view that the disease was spread through the air in what was the prevalent "miasma" theory [5]. Only later in 1866 did Dr. Farr, who had strongly opposed Dr. Snow's findings in 1854, conclude after another cholera outbreak in the Bromley-by-Bow Borough of London, that Dr. Snow was right after all. He determined that infected water with fecal material was indeed the source of spread. Over a decade after the 1854 epidemic, Dr. Farr strongly championed the idea of boiling water to purify it before drinking. This method proved to be the key to stopping the spread of cholera [6].

Today, there is an annual ceremony in London that involves removing the pump handle of the Broad Street pump. This ceremony occurs even though the street is now named Broadwick Street. The pump now stands as a memorial to Dr. Snow's finding and no longer pumps water (Fig. 3.2).

Fig. 3.2 The Broad Street pump, which proved to be the source of the 1854 cholera outbreak. (London Historical Society)

Human Genome Mapping

The human genome program began in 1990 and took over a decade to complete. The final decoding required finding the pairing of 3.3 billion pairs (see Fig. 3.3) [7]. Just one of the positive outcomes of this herculean effort has been the ability to detect the mutation that leads to sickle cell disease and then allow a correction to occur. It is thought that in time, many diseases that affect the human immune system may be cured or prevented by finding a way to fix these mutations. Our understanding of how the four components of DNA, namely adenine, cytosine, guanine, and thymine, will be of enormous help to future medical research.

Remote Sensing and Data Analytics

The latest satellite remote sensing systems and high-powered data analytics could provide powerful new tools to detect the onset of pandemics and contain them.

One such system plans to deploy a network of over a hundred sophisticated satellites that provide imaging from radar, infrared, and visible light spectra.

Fig. 3.3 The first printout of the human genome. (Credit: Russ London, CC BY-SA 3.0, https://commons.wikimedia.org/w/index.php?curid=9923576)

Some would have hyper-spectral capability. This satellite system would be connected to a constellation of high-speed relay satellites that with only a 3-second delay can relay an entire image of the Earth and do so in incredible detail. With some preprocessing and compression of the data, a bank of high-speed processors is able to receive an amazing 27 petabytes of data a day. A petabyte is 1024 trillion bytes of data (2^{50} power).

The most amazing aspect of such networks is not how much near-instantaneous data can be collected. No: it is the sophisticated algorithms that can be devised to monitor patterns of commercial, scientific, medical, governmental, and daily life on planet Earth. One possible form of analysis would be to detect any unusual rise in trips to hospitals and clinics all over the globe. In time, there could well be dozens of algorithms developed to assess medical treatment trends.

Artificial Intelligence and Machine Language

Michael Crichton's thriller novel *Congo* included a scene in which a slightly unhinged General testifies to the U.S. Congress about the need for artificial intelligence (AI) to control all of the millions if not billions of weapon interactions that might take place within less than a second in the next Great War. He explained that human reactions were much too slow to command automated weapon systems, and thus, a transition to AI-controlled systems was required. AI-controlled weapons systems will very likely be rejected by any rational person, but an automatic, global compilation and reporting system on infectious disease is now technically possible and seems worthy of serious consideration. If there were data systems installed in all of the hospitals and clinics in the world that automatically reported instances of infectious diseases, then aggregated, and analyzed these instances within regional and global systems, the outbreak of a serious new infectious disease could be very quickly noted across the medical community. Such a system could likely be designed and implemented for less than a few billion dollars. The COVID-19 pandemic has now cost trillions of dollars and on the order of a million lives as of the end of 2020. Since early containment of a new infectious disease outbreak is the most important preventive measure, the creation of such a system just from an economic and life-saving viewpoint seems almost to be a no-brainer.

The U.S. Center for Disease Control (CDC) established what is known as the National Outbreak Reporting System (NORS) back in 2009. This is a web-based platform where doctors and health officials in the United States would report on disease outbreaks to the CDC database. This includes diseases transmitted by infected persons, animals, or unknown modes. Other countries have similar systems, and many are now interconnected. The NORS system could represent a prototype for a fully global system.

Currently, the World Health Organization (WHO) operates a Public Health Emergency Operations Centre Network (EOC-NET). This provides global outbreak alerts of any new infectious disease and is accomplished through its Strategic Health Operations Centre (SHOC). In addition, the WHO also seeks to promote international collaboration and cooperative response during pandemics. This system, however, is dependent on national health services promptly reporting outbreaks under the mandates of the International Heath Regulations. As was shown in the case of the COVID-19 pandemic, some countries can be slow to report an outbreak and instead seek to curtail it locally before providing an international alert of the health danger.

The WHO's Strategic Health Operations Centre is only as effective as the national health alerts it receives. Thus, a new health alert system based on satellite sensing and data analytics could provide much earlier alerts of viruses and infectious diseases. This is because outbreak detection processes would not be filtered through governmental health systems before alerts are given.

Clearly, not all territories of the world have rapid pandemic alert capabilities, nor do they have the ability to respond effectively after the outbreak is detected. Thus, many countries, through grants, should be funded to be equipped with more ventilators, together with better local medical diagnostic and communications equipment to participate in a comprehensive pandemic alert system. Countries without extensive medical capabilities and infrastructure should at least be scanned by high capability remote sensing satellite systems and smart data analytic systems. A week of advance notice might dramatically lower the impact of a possible pandemic from severe to mild.

How Space Systems Can Help Cope with Pandemics

Space systems have come a long way since the launch of Sputnik in 1957 some 65 years ago. An in-depth review of how satellite and space systems can help out with pandemics has been prepared by the International Space University. Major conclusions from this study's Executive Summary are quoted below:

- Integrate Earth Observation and Global Information System (GIS) data with crowd-sourced data to monitor severity and spread of infection.
- Provide education and training for accurate data collection and analysis for health personnel, rapid responders, and decision-makers.
- Track supply chain networks to counter the impacts of pandemics by identifying safe and viable transportation routes to deliver critical medical and subsistence supplies in remote or quarantined regions.
- Maintain and manage essential commercial trade for economically resilient communities.
- Tracking animal and human population interactions for disease transmissions and spread, and monitoring environmental factors predictive of disease outbreaks, such as measuring air quality or monitoring flood regions.
- Create predictive risk models to prevent and prepare for disease outbreaks [8].

The various ways to use broadband mobile communications to support the detection, treatment, and logistical issues associated with a pandemic should be clear to most people. What is not widely known is that broadband 5G systems (or even narrower-band 3G or 4G systems) cover only about 8–9% or so of the Earth's total surface. The new high-speed Internet-in-the-sky satellite systems for the most part cover 100% of the entire world.

The Internet constellations will allow remarkable new medical systems for networking, detecting disease outbreaks, and coping with logistical problems. The world's various planned and ongoing space navigation systems, such as the Global Positioning Satellite (GPS), the Russian Glonass System, the Chinese Bei Dou System, the European Galileo System, the Indian Region Navigational System, and the Japanese Quasi-Zenith System will provide major capabilities in navigation, automated transport, Internet synchronization, and more. New capabilities that might appear in another two decades include automated medical kiosks and automated transportation systems on the ground, in the oceans, and in the sky. These will be able to navigate the world without fear of virus infections.

Alien Pathogens?

It is possible that a new and more deadly virus could come from beyond Earth. In the years to come, there will be more and more space travel across the solar system. Eventually, perhaps, there will be efforts to industrialize many types of off-world activities. The possibility must be faced that these activities could over time allow a deadly pathogen to come back from outer space. Today, space scientists tend to believe that life beyond Earth is most likely to exist where there is a supply of water in liquid form. This, as far as we know, designates Mars, Europa, and Enceladus as the most likely source of life outside of Earth within the realm of our solar system.

One of the frustrating problems that exists today is that there is not even a clear definition of the words "planetary defense." The author has worked on several projects related to the subject, including the *Handbook on Cosmic Hazards and Planetary Defense*, as well as the book *Planetary Defense*. (Both published by Springer.) In these books, "planetary defense" was used in a broad sense of defending against comets, asteroids, solar flares, orbital space debris, cosmic rays, antimatter, and off-world infections that might come to Earth.

If one reads the report on Planetary Protection Policy by the U.S. National Academies of Science, Engineering and Medicine, one will learn that within NASA and the COSPAR Panel on Planetary Protection, there is a very narrow definition of this term. The official definition of "planetary protection" or "planetary defense" in the National Academy's 2018 report addresses only the hazards of cosmic infection from space to Earth, or the reverse problem of Earth diseases spreading out into space, in such a way that it might confuse our understanding of the origins of life. The precise definition reads as follows:

> Planetary protection is the practice of protecting solar system bodies (i.e., planets, moons, comets, and asteroids) from contamination by Earth life in order to preserve the opportunity for scientific studies at those destinations relating to the origins of life and/or prebiotic chemical evolution, and protecting Earth's inhabitants and environment from harm that could be caused by possible extraterrestrial life forms [9].

The National Academy policy review group recommends that an interdisciplinary advisory group be established to evaluate alien life forms. This group would include specialists from a broad range of disciplines.

The global framework for undertaking this "planetary protection" against celestial infection is provided under the U.N. Outer Space Treaty of 1967. Article IX states in part:

> In the exploration and use of outer space, including the moon and other celestial bodies, States Parties to the Treaty shall be guided by the principle of co-operation and mutual assistance and shall conduct all their activities in outer space, including the moon and other celestial bodies, with due regard to the corresponding interests of all other States Parties to the Treaty. States Parties to

the Treaty shall pursue studies of outer space, including the moon and other celestial bodies, and conduct exploration of them so as to avoid their harmful contamination and also adverse changes in the environment of the Earth resulting from the introduction of extraterrestrial matter and, where necessary, shall adopt appropriate measures for this purpose... [10]

Article XI also states in part that:

States Parties to the Treaty shall bear international responsibility for national activities in outer space, including the moon and other celestial bodies, whether such activities are carried on by governmental agencies or by nongovernmental entities, and for assuring that national activities are carried out in conformity with the provisions set forth in the present Treaty... [10]

At the international level, the Committee on Space Research (COSPAR) has established a Panel on Planetary Protection to coordinate a global Planetary Protection Policy. Space agencies engaged in deep-space exploration, such as NASA, CSA, JAXA, ESA, ISRO, the UAE Space Agency, and the Chinese National Space Agency, as well as private commercial organizations that envision such activities as space mining of asteroids, are asked to develop their own protective policies consistent with the COSPAR Planetary Protection Policy [11]. The problem is that as more space agencies and other actors seek to engage in these space activities, protecting against infection of or from celestial bodies will become increasingly difficult.

Why an Integrated Approach Is Necessary

One could argue that there are only tenuous links between medical science, solar storms, asteroid and comet strikes, orbital space debris, natural disasters, overpopulation, species endangerment, climate change, and artificial intelligence. Even within space agencies such as NASA, the scientific experts who work on orbital space debris, solar storms and CMEs, and asteroid and comet strikes are in totally different units of the agency. But the arguments that try to separate these hazards by a narrow discipline focus are in many ways wrong-minded. True understanding of the problem of human pathogens and their effective treatment needs a multidisciplinary or interdisciplinary approach.

As Dr. Jared Diamond put it: "This process of transforming animal microbes into human pathogens is accelerating today, but it is not new. It began…when we first cleared wildlife habitat to make way for crops and yoked wild animals into servitude" [12]. Dr. Diamond feels that a broader perspective is needed to research and understand viral infections, zoonotic diseases, or even the threat of viruses from outer space. Whatever the field of research, to narrow the view can be a mistake.

The power of massive amounts of "big data," along with new algorithms and ways of processing them, may help us discover important relationships between habitat changes and infectious pathogens. Whatever the case, interdisciplinary studies and connections across small-scale and large-scale systems will likely prove to be the key to major scientific advances of the future.

Pioneering efforts in this regard include organizations such as the Santa Fe Institute. This innovative U.S. program brings together Nobel Laureates and leading scholars and researchers in economics, astrophysics, chemistry, mathematics, and medicine to foster interdisciplinary interaction. Some have carried out research and written about such topics as complexity, nonlinear mathematics, fractals and complex adaptive systems, economics, microbiology, medicine, statistics, chemistry, physics, and information sciences. They believe that interdisciplinary interaction can reveal totally new causal relationships. They have found that this interdisciplinary perspective can apply to subjects from market behavior to the nature of geographic coastlines and from the structure of the universe to the structure of viruses.

Conclusions

Our experience with MERS, SARS, Ebola, AIDS/HIV, and now COVID-19 shows how risks to humanity are occurring at an accelerating rate. COVID-19 has infected many tens of millions of people and may ultimately end up taking millions of lives and crippling the global economy. Yet today, there are walls and silos everywhere. Specialization of study seems intent on slicing up the world into smaller and smaller parts, risking our overall understanding of many problems the world faces today. As mentioned previously, NASA has different units working on asteroid and comet strikes, solar storms, and orbital space debris. Even medical research tends to be highly specialized as well. These separations exist across national boundaries, governments, universities, and private enterprises. Of course, specialized research is needed. Yet a much more integrated, interdisciplinary, and international approach to modern threats could pay huge dividends. There is no need to abandoned narrow and targeted research, but this also doesn't mean that we cannot have interdisciplinary research teams for advanced researchers.

We need international teams of scientists, engineers, ecologists, medical doctors, astrophysicists, economists, political experts, and innovative thinkers to work together in today's advanced global research environment. An essential step is flexible data sharing. Part of the reason for the division of technologies and separate fields of study in some cases is for commercial profit. The push to develop dozens of different vaccines for COVID, for instance, is motivated by Big Pharma competition. Competition for new ideas and innovations is good. However, artificially structured separation of innovation for profit, particularly in the field of medical research, is often bad and potentially dangerous.

Such a global effort might be initially staffed from the World Health Organization, the Food and Agricultural Organization, the World Meteorological Organization, the UN Environmental Program, the UN Office of Outer Space Affairs, the world space agencies, the UN Office of Disarmament Affairs, the UNESCO science program, and perhaps industry professionals. This worldwide research group would be charged with exploring the dimensions and causes of various global threats to the sustainability of Earth. The initial sustainability research team should perhaps be capped in size at perhaps some 50–100 people. This "thought squad" could help create a global emergency civil defense research system. The group might also be linked together trained experts who are tasks with responding to global disasters and threats. It would thus also be linked to rescue workers and medical response teams, as well as military units trained for emergency response. Pandemics are just one of the threats that could be reduced or responded to more effectively in this way. The specifics of how a better global approach might be taken are addressed in Chap. 14. This is just to indicate that there is a need for a worldwide effort to make the world resilient against significant longer term threats. An International Agreement to work together toward global sustainability would be a key step in that direction. New uses of space and data systems need to be a part of this global plan.

References

1. Malaysia identifies new Covid-19 strain similar to one found in three other countries. Strait Times, December 3, 2020. https://www.straitstimes.com/asia/se-asia/malaysia-has-identified-new-covid-19-strain-similar-to-one-found-in-3-other-nations
2. Vital, J.: Destruction of habitat and loss of biodiversity are creating the perfect conditions for diseases like COVID-19 to emerge. Popul. Connect. **52**(2), 15 (2020)

3. History's Most Deadly Pandemics, from the Antonine Plague to Covid-19" Discoverer. May 14, 2020, https://www.discovermagazine.com/health/historys-most-deadly-pandemics-from-the-antonine-plague-to-covid-19

4. Louis Pasteur: Biography (1822–1895). https://www.biography.com/scientist/louis-pasteur

5. Begum, F.: Mapping Disease: John Snow and Cholera. https://www.rcseng.ac.uk/library-and-publications/library/blog/mapping-disease-john-snow-and-cholera/

6. Bingham, P., Verlander, N.Q., Cheal, M.J.: John Snow, William Farr and the 1849 Outbreak of Cholera. https://pubmed.ncbi.nlm.nih.gov/15313591/. Last accessed 25 Sept 2020

7. The Human Genome Project. https://www.genome.gov/human-genome-project. Last accessed 25 Sept 2020

8. The Role of Space During Pandemics. International Space University, 2020. https://isulibrary.isunet.edu/doc_num.php?explnum_id=1766

9. Committee on the Review of Planetary Protection Policy Development Processes: Review and Assessment of Planetary Protection Policy Development Processes. National Academy of Sciences, Engineering, and Medicine, Washington, DC (2018)

10. United Nations Outer Space Treaty of 1967. https://www.unoosa.org/oosa/en/ourwork/spacelaw/treaties/outerspacetreaty.html. Last accessed 30 Sept 2020

11. Op cit COSPAR Committee on Planetary Protection, 2018

12. Shah, S.: Think exotic animals are to blame for the coronavirus? Think again. The Nation, February 18, 2020

4

Pollution

Pollution is nothing but the resources that we are not harvesting. We allow them to disperse because we've been ignorant of their value.

–R. Buckminster Fuller

Large quantities of plastic are continuously ending up in our oceans. Microplastics enter the diet of fish, shellfish, and birds, as well as our own food.

–The Norwegian Geotechnical Institute

Introduction

Perhaps everyone will recall the four horsemen of the biblical apocalypse: conquest, famine, war, and death. The four scourges that constitute the modern apocalypse are as follows: overpopulation; endangerment—or extinction—of plants, animal life, and vital resources; climate change; and pollution. Our continuing failure to mount a successful campaign against these four problems will create truly dangerous problems for humanity in the remaining decades of the twenty-first century.

We will lose a countless number of species that now exist on Earth. This will primarily happen due to the effects of climate change. There will also be losses driven by the unsustainable growth of the human population. In additional, there will be losses due to overconsumption and our disposal economy. Unless significant progress is made now, the problem will get progressively worse, to the point where one can expect the arrival of a fifth modern horseman: mass extinction.

There have been five mass extinctions events in the history of the world to date. Most of these have come from major changes in the average temperature on the planet. The last one—a huge asteroid that smashed into Earth 66 million years ago—will be addressed in Chap. 6. According to scientific and geological studies, 99.5% of all the species that ever existed on Earth are now extinct [1]. Without reform, the next extinction event, sometimes called the

J. N. Pelton, *Space Systems and Sustainability*, https://doi.org/10.1007/978-3-030-75735-9_4

Anthropocene extinction event, will wipe out most life forms on Earth—humans included. What an inconvenience that would be.

Environmental Initiatives

This chapter focuses on the hazards of global pollution in the atmosphere, on land, and in the seas, and what to do about it. Of course, to truly address climate change, there must also be progress on population control, nature conservation and habitat protection, and global pollution, especially in the oceans. Overall, progress will only be made by seeing the four modern horsemen of the apocalypse as a whole and launching an integrated attack against them.

Global cooperation is essential, and it must include strict prohibitions, penalties, and especially incentives to do the right thing. The UN Environmental Program (UNEP) is underequipped, undermanned, and underfinanced. The UNEP seven programs to address the environment are entirely worthwhile efforts. Yet such underfunded programs are unable to steer a world economy of more than $80 trillion dollars (US). Massive reforms are needed. Jet aircraft needs to be converted into electric propulsion. Motor vehicles need to be converted into electric as well. Carbon-based fuels for heating, transportation, and other uses need to be eliminated from the world's ecosystem. Nowhere do the regulatory or economic controls currently exist to achieve such massive and long-term reforms. Many of the needed reforms will take not years, but decades to achieve.

They say it takes at least 10 miles (16 km) or more to turn an aircraft carrier around. Turning around the world economy to make it cleaner and greener in the next two or three decades would be a miraculous achievement. Making the world truly green would be a much more monumental achievement. But this is an even bigger task than turning a large fleet of aircraft carriers around. Nowhere today are there the tools to do this within the U.N. system or other entities such as the European Union or G20 countries. The one slim hope might be within the World Economic Council, which holds its meetings in Davos, Switzerland. This group is truly seeking to understand the world's macroeconomic programs and its most dangerous modern dangers. The WEC has established some 80 study committees to address many of the world's greatest challenges, including those related to the environment, pollution, and space systems.

It is hoped that eco-conscious nations and their governments might be willing to take on the most difficult of the world's pollution problems. Such

lead countries would adopt strict environmental laws with suitable legal penalties, fines, as well as financial incentives for reforms, and then apply pressure
on more and more countries to follow suit with parallel laws and enforcement
provisions.

The modern four horsemen will require different forms of legal and regulatory controls as well as financial incentives at the local, state, and global level.
There can be several ways forward. The process outlined under the Global
Sustainability Treaty in Chap. 14 and the review process to identify global
threats outlined there could play a role. Another option might be for well-
known environmental reform groups to draft an environmental manifesto
with a comprehensive listing of proposed legislative provisions, which would
create model laws to halt the most dangerous forms of pollution.

Such a model law would seek to ban harmful pollutants of the air, land, and
sea, reward power plant conversion from coal, set targets for electric jet propulsion for aircraft, and establish goals for restoration of the oceans. This
would identify year-by-year levels for reduced ocean acidity and set thermal
goals for ocean temperatures, among dozens of other provisions. Clearly,
enforcement processes and penalties will be required, but the key may be
financial incentives that reward actions to move to cleaner and greener
programs.

The critical next step after this model law is drafted would be for a core
group of leading nations to adopt the provisions as enforceable national law.
The difficult part of the process would be getting a critical mass of nations to
agree to move forward in tandem. After all, these environmental laws will
alter corporate behavior and enforcement powers. It could be that recalcitrant
global industries could try to whipsaw the process to play one country against
another.

It is a good goal to get perhaps the top 10 industrial nations on board. Stage
2 would be to receive commitment from the next top 20 to 25 largest industrial countries. At this stage, the biggest global polluters would largely be on
board. The additional follow-on effort would put the squeeze on countries
that might serve as "flag of convenience" polluters. This might also involve
enticing incentives.

For years, there has been an effort to address global pollution from the top
down through international treaties and efforts, such as the Paris Accord of
2015. Yet international "goals" are just that. They are neither mandatory nor
enforceable. Only national legal provisions backed with both incentives and
penalties can get results in truly meaningful ways.

The history of national legislation action in the United States to create new
mileage standards for automobiles is perhaps an instructive example. The

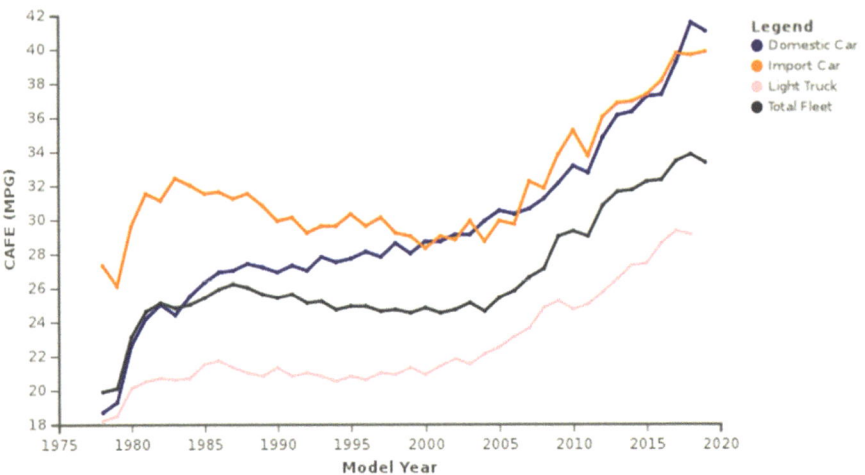

Fig. 4.1 The remarkable success of the CAFÉ legally mandated increase in vehicle fuel efficiency. (Courtesy of the U.S. Highway Traffic Safety Administration)

industry opposed such standards and claimed they were simply not technically possible. Yet faced with the reality of deadlines, the industry found ways to meet mandated standards for the Corporate Average Fuel Efficiency (CAFÉ) legislation. During the period between 1978 and 2019, the average fuel efficiency for cars went from 18 miles/gallon to over 40 miles/gallon [2] (See Fig. 4.1).

Perhaps this effort might also be supplemented by the initiative of an environmentally friendly billionaire. This individual might create an equivalent of the Nobel Prize for national governments that adopted and enforced the new environmental laws with the greatest rigor. Such a "Nobel Prize for Environmental Improvement" would be largely based on detailed satellite and high-altitude platform systems data and ground- and ocean-based sampling systems. Each country would be monitored based on remote sensing imaging, river and lake sampling, and data relayed from buoys near national ocean shorelines.

There could be two environmental scores, with the winner in each category receiving the top award. Each country would then be ranked on such indices as the loss or improvement of tree canopy, the level of greenhouse gas emissions and output from coal-fired plants and refineries, pollution of seas and oceans, particularly, in proximity to national shore lines, acidity of rain levels, and other pollution indicators. The ranking system would go from first to last, with the least polluted countries winning awards for first, second, and third. The second score would be for percentage of improvement from the previous year. Twenty-five percent of the improvement score would come from strengthening of environmental legislation and regulations that reduce pollution levels and limit climate change.

The precision of satellite sensing systems could give an incredibly detailed profile of the Earth's pollution levels broken down by nation. It could show with precision where progress was being made and especially areas where deterioration was occurring. At a lower level, there might be a special award for organizations, national governments, or corporations that contribute the most to reducing ocean pollution, with special categories for thermal pollution, oil spill pollution, plastic pollution, and preservation of ocean life and carbon sink systems.

There might be other prizes for antipollution technologies, systems, and policies. It would be wonderful to see countries competing for the "Clean Earth" awards, incentivized by a $10 million prize and medal.

Types of Global Pollution

Some types of pollution, including poisons, radiological wastes, and elements of warfare, such as nerve gases, have dramatic effects. These type of pollutants represent instruments of immediate life and death. Other forms of pollution require much longer for their impact to be felt. Sometimes pollution results in the destruction of plant life that ultimately leads to starvation, or perhaps it results in a mutation that leads to stillborn or misshapen infants. One of the most serious long-term effects could be the increase in temperature and acidity of the ocean. These effects can kill coral reefs and cyanobacteria. The process may take decades, but the ultimate outcomes might well be direr than a poison with immediate effects. Today, hyperspectral remote sensing satellites and thermal sensors on remote sensing satellites can show the deterioration of the coral reefs with great precision and sound alarms well before all coral systems have completely died. In the past, satellite systems did not have sufficient resolution, worked in too wide a band range or only used a particular type of active or passive system, but this is no longer the case [3].

Some of the most powerful tools available to detect pollution in the ocean today are the NASA Aquarius satellite and the European Space Agency's Soil Moisture and Ocean Salinity (SMOS) satellites, although they are now nearing their end of life. These satellites map ocean salinity, which indicates alkalinity or the ability to neutralize acidity. Yet another more recent tool is NASA's Carbon Observatory (OCO-2), launched in December 2014, which can measure carbon buildup and acidity directly.

At least a quarter of carbon dioxide in the air is absorbed by algae and plankton or is absorbed directly into the oceans. Measurements by satellites have shown areas where this acidic buildup is taking a significant toll on sea life, such as in the Polar Regions and the Bay of Bengal.

Satellites are also key for measuring the steady increase in ocean temperature. The combined effects of thermal heating, carbon-based acid buildup, oil spills, and plastic poisoning are a four-way whammy to oceans and sea life [4] (See Fig. 4.2).

Chemicals designed as insecticides can have disastrous long-term effects on plants, animals, and ultimately, humans. In this case, one can think of DDT, whose adverse effects took decades to determine. Aerosols and other pollutants contributing to the ozone hole in the Polar Regions have led to deadly mutations among frogs and other species in Southern climes.

Radiological dangers can also come in various forms. A reactor at a nuclear power station can meltdown rapidly and create very high, lethal radiation levels in a short period of time. Faulty disposal of nuclear waste and low levels

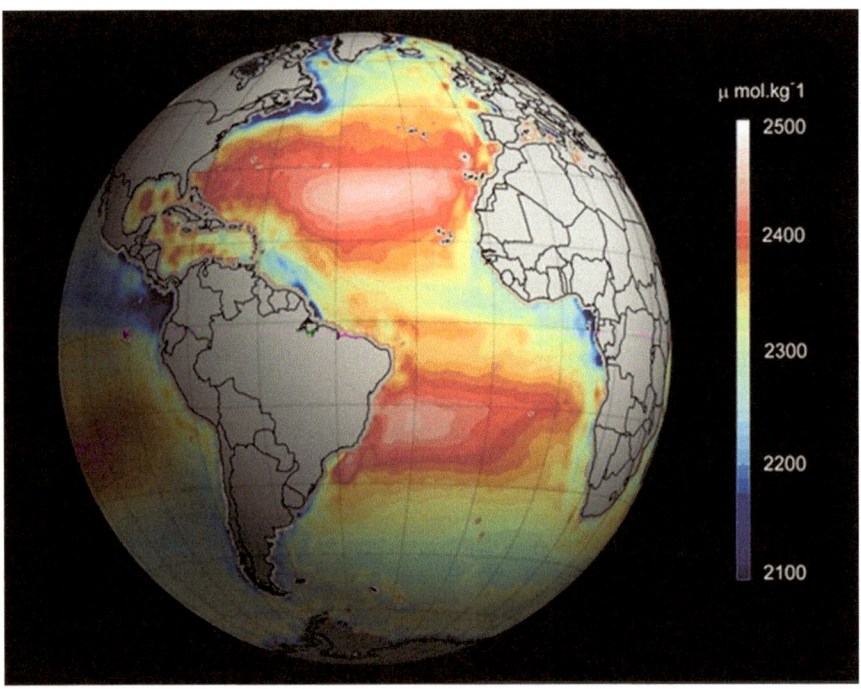

Fig. 4.2 Satellite imaging reveals the dangers of ocean acidity increases. (Credit: ESA)

of radiation leakage can take years for the negative effects to register but bring death all the same.

There are also different forms of pollution that are generational in impact. These include noxious materials such as defoliants used in warfare, like Agent Orange, which was employed in the Vietnam War. These chemical weapons can lead to birth defects and debilitating longer-term effects on humans.

Finally, greenhouse gases build up and can lead to global warming and climate change. A rise in temperature around the world causes the thawing of peat fields in Siberia, Alaska, and Canada, which release billions of tons of methane into the atmosphere.

Food and Agriculture

Human activity is creating ever great levels of pollution and contamination of crops and vegetation, whether grown for humans or not. The extent of this problem has just begun to be recognized. The recently detected problem of microplastic contamination will be discussed below, in addition to the problem of genetically modified organisms.

Modern agriculture has been scientifically engineered to be more productive per acre of cultivated fields. Beyond chemical fertilizers, there are now "smart tractors" on large agribusiness farms that use remote sensing satellite data to cover huge tracts of farmland. On a square meter by square meter basis, these machines precisely control the amount of fertilizer, seeds, insecticides, and water to place on the ground. These smart farming processes allow automated farm businesses to grow more wheat, corn, soybeans, potatoes, onions, and other crops.

Over time, however, farmlands around the world are being polluted from more and more sources. Insecticides, which have become known as persistent organic pollutants (POPs), are perhaps the biggest single source of concern. Aldrin, chlordane, DDT, dieldrin, endrin, heptachlor, hexachlorobenzene, mirex, and toxaphene are some of the more notorious insect poisons. Some of these are now considered carcinogens that have been labeled as too dangerous to use on food crops. Chemical fertilizers still used on farms and lawns run off and ultimately pollute rivers, lakes, bays, seas, and oceans [5].

Again, satellite technology and data analytics can provide new capabilities. High-speed data analytics can be employed to detect particular spectral signatures that indicate the presence of the various pollutants and poisons. Also, there is the possibility to create global systems that provide very rapid updates not only about the existence of poisons and pollutants but also the dispersion and currents of those poisons and pollutants into rivers, seas, and oceans.

Hazardous chemicals from industrial factories, plus radiological waste from nuclear power plants, hospitals, and medical research centers, also often find their way into water runoff and become part of the pollution cycle. Over time, pollutants build up in the land where crops grow and in the rivers and seas where fish are caught. Some ocean fish have now been ruled dangerous to eat in large portions because of high levels of mercury content, including: King mackerel, marlin, orange roughy, shark, swordfish, tilefish, ahi tuna, and big-eye tuna. The buildup of mercury in fish is expected to increase over time [6].

In many countries, there are now agencies such as the Environmental Protection Agency (EPA) that seek to establish levels of purity for food and water, protecting against dangerous pollutants that would get into the food supply. One of the main concerns is contaminated fish. There have been attempts to ban fishing in contaminated waters, but sometimes fishing boats disregard limits and engage in illegal fishing, bringing fish to market that are not safe to eat. The greatest danger in food and water consumption today remains tainted supplies that can cause things like E. coli poisoning, as discussed below.

Again, space tools such as hyperspectral sensing, data analytics, automatic identification systems, global RF mapping systems, and especially artificially intelligent data analytic systems can spot polluters, track the source of pollutants, and much more. Traditionally, governmental operators have used many of the space systems for applications such as weather monitoring, geological mapping, and military or policing applications. They have not designed these systems for environmental applications. Private operators have geared the use of their systems for commercial applications with industrial, farming, or forest management applications, and also have not designed their data analytics for environmental infractions.

Genetically Modified Organisms (GMOs) and Natural Genetic Mutations

Companies all over the world have food processing and production laboratories filled with organic chemists. They are studying ways to make their food products more productive, plentiful, and profitable. Genetic modification has allowed food to stay fresh longer, resist more types of insects, and grow more quickly to maturity or to a larger size.

There are many who worry whether such food products might lead to cancer or other problems, including increased antibiotic resistance, allergens, unnatural nutritional changes, and even mild levels of toxicity and possible harmful genetic mutations. There is also concern that so-called "out-crossings" will occur, where a GMO food product invades other plant species. It has been noted that it took years to discover some of the harmful effects of DDT on the human body, but that GMO food products are developed quickly and then introduced into the world's food supply.

One might be reassured by the fact that there are only a few genetically modified foods grown in the United States that have been approved by the U.S. Department of Agriculture, the Food and Drug Administration, and the Environmental Protection Agency. Yet this does not tell the entire story. The question is not how many GMOs have been approved, but how prevalent are they? "Less than a dozen genetically modified crops are grown in the United States, but they often make up an overwhelming majority of the crop grown. More than 90% of soybeans, corn and sugar beets planted in 2018 were genetically modified." [7]

Under the Trump Administration, a new $7.5 million "Feed Your Mind" campaign was launched to systematically provide information about the science behind genetically modified organisms (GMOs), with the intent to bolster the general public's confidence that these food products are safe to purchase and eat [7].

Microplastic Poisoning

Microplastic particles are a form of pollution created when huge amounts of plastics floating in the ocean and seas are broken down by the sun and ocean wave action. This process has been going on long enough now that fish and other water life have systematically begun to consume significant amounts of microplastics.

There are many ways that plastics find their way into the food chain, including via storm sewers, water mains, and various drainage systems. Many billions of bags and plastic water bottles find their final destination in the oceans. Beyond microplastics, the plastic bags ensnare the gills of fish and trap ocean wildlife (Fig. 4.3).

The mounting concern over ocean-based plastic pollution and the lack of a systematic way to recycle disposable plastic bags and containers have led to the banning of plastics for such purposes in many jurisdictions. To date, Bangladesh, China, the Congo, Italy, Kenya, Rwanda, and South Africa have

Fig. 4.3 A sea turtle trapped in plastic straw. Many species are now considered to be endangered. (Courtesy of Francis Periz, photographer, and the Blue Ocean Conservancy)

banned disposable plastic bags. In the USA, two states, North Carolina and Hawaii, have banned plastic bags, while other jurisdictions have imposed plastic bag fees. This has altered consumer practice so that many consumers are now routinely using cloth tote bags instead. On a 2019 trip to Cape Town, South Africa, this author noticed that many disposable plastic bags were still in use despite the ban, but large chain stores such as Woolworth's were only selling large, permanent tote bags.

Despite these efforts, plastic waste continues to grow. Trash processing plants have found that thin plastics tend to jam their automated systems, which are designed to process a combined or single stream of recyclable waste. Widely used polystyrene materials (more commonly known as the trade-marked Styrofoam) cannot be recycled effectively. Many countries that once accepted and recycled or disposed of plastic waste, such as China, Malaysia, Vietnam, Thailand, and others, have now banned this practice. Some Pacific Island countries are feeling the impact on fish, shellfish, birds, and other forms of sea life, as well as on littered beaches.

Systems are now being designed to capture ocean-based plastic waste. One such design would encircle a large ocean area strewn with plastics with a floating, mile-long system. This plastic capture system is then tightened into a noose to trap tons of plastic waste. This is just one of many efforts to eventually remove a huge amount of plastic flotsam from the sea. Other campaigns are designed simply to get consumers to pledge not to use plastic containers or bags. Another effective strategy is charging consumers for plastic bags.

Other Forms of Ocean and Water Pollution

There are many more ways in which ocean water can become polluted. Oil spills from tankers and accidents involving oil drilling, particularly in the case of deep-water platforms that are disrupted or destroyed, are one of the most serious sources. Some of the latest industry ideas involve mining of metals and other resources from the sea floor, and others are looking to process suspended particles in the ocean waters to reclaim them as metals or other types of resources. These operations can cause direct pollution from the reclamation process and disrupt the diets and habitats of sea life.

As noted in the subsequent solutions section, cyanobacteria and phytoplankton produce the great majority of oxygen on the planet. These photosynthesis-driven organisms produce the purified air that we breathe in larger quantities than the rain forests. Industrial processes cannot be allowed to disrupt these oxygenators. Too often, regulatory restrictions respond to disastrous industrial activities after they have occurred, rather than preventing ecological mishaps beforehand. Protection of the ecosystems that supply oxygen to all life must be a top priority.

Water pollution must be addressed at two levels. One level involves protection of oceans, rivers, and streams so they are pure enough to sustain fish, birds, beavers, otters, and more. The other level involves creating clean water reservoirs protected from infectious diseases and treated to keep everyone who drinks it healthy. As is clear from the earlier chapter on pandemics, water can be easily polluted by feces and other organic matter. Water after a disaster such as a hurricane or tropical storm can end up carrying diseases and parasites to local populations [8].

Impure water is a major source of disease. Many parts of the world have polluted water supplies, and this problem is greatly amplified by rapidly growing human population, climate change, desertification, and urbanization. New methods of water purification and desalinization systems are worthy areas for research. Unfortunately, many of the ideas under development are stopgap measures and not longer-term, sustainable answers.

Radiation and Nuclear Poisoning

There are about 150 nuclear reactors operating in the United States. This number includes just over 50 research reactors, largely at universities. The other hundred reactors are operated by private nuclear power providers across the country. In a number of cases, there are three reactors clustered at one site

U.S. Operating Commercial Nuclear Power Reactors

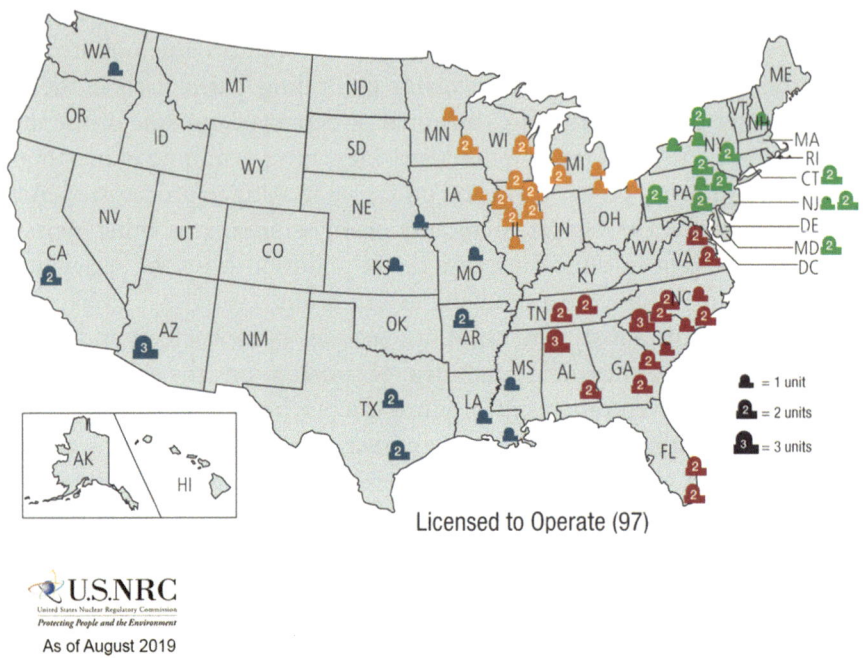

Licensed to Operate (97)

U.S.NRC
United States Nuclear Regulatory Commission
Protecting People and the Environment
As of August 2019

Fig. 4.4 The 97 nuclear power plants operating in the United States, excluding the more than 50 research reactors. (Courtesy of the US Nuclear Regulatory Commission)

(see Fig. 4.4). Often, these sites are located in a coastal location, which puts them at risk of a disastrous tsunami and/or earthquake event similar to that which occurred in 2011 in Fukushima Daiichi, Japan. The three reactors located there were not designed to be able to withstand the 15-m tsunami wave that destroyed all three reactors and spread radiation over a wide area [9].

Many thousands of reactors are now operational around the world. In most instances, there are no agreed permanent storage locations or systematic oversight for disposal of toxic waste materials. These are all at risk of potential reactor meltdowns and production of more radioactive waste products. Beyond this, there are still thousands of nuclear devices stockpiled for warfare purposes. Additionally, there are a very large number of radioactive materials used for medical and other industrial purposes. So many dangers are associated with the production, transport, and safe storage of these products. This includes the possibility that the materials might be stolen and used to create a so-called "dirty bomb."

Solutions and the U.N. Sustainable Development Goals

The U.N. 17 Sustainable Development Goals set pollution mitigation objectives within at least eight goals. They are as follows:

- Goal 3: Good health and well-being
- Goal 6: Clean water and sanitation
- Goal 7: Affordable and clean energy
- Goal 11: Sustainable cities and communities
- Goal 12: Responsible consumption and production
- Goal 13: Climate action
- Goal 14: Life below water
- Goal 15: Life on land

This approach has at least two problems. First, there is no focused or unified effort to reduce global pollution. Second, other U.N. goals could and probably do work *against* containing pollution. The specific goals that might run counter to this aim include:

- Goal 1: No poverty
- Goal 2: Zero hunger
- Goal 8: Decent work and economic growth
- Goal 9: Industry innovation and infrastructure

Part of the problem is the missing goal in the U.N. list: to stabilize and gradually reduce the rate of human population growth. This is the biggest driving force that gives rise to industrial pollution, hunger, poverty, unmet demand for education and health care, insufficient energy, etc. It is the huge elephant in the room that everyone chooses to ignore. Nevertheless, the U.N. Sustainable Development Goals are indeed a useful way to analyze global problems and set useful targets for advancement in many areas.

Coal and carbon-based energy systems ultimately pollute and harm the atmosphere, land, and water. All of these need to be replaced in future decades. There is a need for clean, nonpolluting electrical power systems, automotive vehicles, trains, planes, and heating and cooling systems and more. Such changes require a unified industrial, business, and financial revolution. Whole industries will have to be reconstituted. The oil, energy, automotive, air transportation, and other industries will have to be reshaped. Only a massive commitment to refocus trillions of dollars will suffice. Currently, there is no mechanism by which such a change could be implemented. Only dramatic legislative reform backed by a willing community of industrialists could make this happen. Today's modest and incremental changes are much too slow. A few factors such as the increasingly high cost of coal and the more cost-effective wind and solar energy systems help, but incentives to "go electric" must be increased. Satellite remote sensing and telecommunications systems and data analytics have the power to detect the shape, form, and location of the biggest problems and create the tools needed for global enforcement programs.

Most of the pollution associated with climate change is linked to not capturing and recycling hydrocarbon-based gases, which are associated with heating and cooling of buildings and transportation systems. The subsequent trapping of greenhouse gases (GHG's) is heating up the Earth's mean temperature, in turn releasing massive amounts of frozen GHGs trapped in polar regions—especially methane trapped in peat bogs in Siberia, Northern Canada, Alaska, etc.

Phytoplankton, cyanobacteria, and other types of bacteria that live in the sea water (think of red, brown, blue, and green algae) are the essential engines of oxygen production on Earth. Cyanobacteria and phytoplankton first evolved on Earth 3 billion years ago, and oxygen-producing bacteria are thought to have evolved about 2.4 billion years ago. Today, microscopic descendants of these oxygen-generating bacteria have been supplying this life-giving element for animal life to breath since that time. Some scientists believe they actually produce 50% to as much as 80% of the world's oxygen today. Just as importantly, they use methane and carbon dioxide to produce the oxygen. Today, unfortunately, there is increasing acidification of the ocean. Recent scientific studies have concluded that in response to this increased acidity, many species of phytoplankton and cyanobacteria may die out, while others may migrate to less acidic waters. It may be that acidification of the ocean could represent the most important form of global pollution of all [10].

At the Eden Center in Cornwall, England, there is the massive 8-meter-tall "Infinity Blue" ceramic sculpture of a "cyanobacteria stack." This monumental artifact was created by Azusa Murakami and Alexander Groves for the Eden Center. The sculpture represents a stromatolite, or a massive stack of fossilized cyanobacteria (see Fig. 4.5). It features "cyan dots" that represent cyanobacteria.

There could come a time when oxygen farms inhabited by these bacteria may be necessary. Indeed, this type of oxygen farms and "carbon sinks" might

Fig. 4.5 The Eden Center's 8-meter-high ceramic sculpture of a cyanobacteria stack that produces oxygen and absorbs carbon. (Graphic provided by the author)

also need to be sent into space to generate oxygen for space colonies and process carbon dioxide waste.

The great diversity of pollutants and their impact are difficult to summarize. Current pollution controls address only a small number of the global dangers that exist today. The U.N. Environmental Programme and its International Programme on Climate Change, its programs on Environmental Rights and Governance, and its International Programme on Biodiversity and Ecosystem Services (IPBES) are currently no match for the world's industrial production systems, which are churning out gigatons of plastics, chemicals, and carbon- and oil-based products every day. The truth is that these industries are very efficiently polluting the world at an ever greater pace. For years, scientists were aware of the harmful effects of pollution on the land, in the air, and in rivers and streams. But only in the past two decades through remote sensing satellite observation was the extent of dangers from pollution on our oceans revealed [11].

Conclusions

We must employ the latest space-based communications, remote sensing and monitoring capabilities, plus super-fast data analytics to help observe and combat climate change. In many cases, this might be best achieved through public-private partnerships. It is only through such modern space systems that we can reach more effective global enforcement of environmental pollution laws. Of course, the key is not only improving enforcement but also making better efforts to prevent the disposal of waste and harmful materials. As engineer and futurist Buckminster Fuller wisely observed, many pollutants, if recovered, reprocessed, and recycled, can be made into useful products and can actually become assets.

The latest space tools can change how the Earth is perceived. Hyperspectral remote sensing satellites can detect different types of plastic pollution in the ocean or understand how quickly a coral reef's health is changing. Forest fires and dangers to vital cyanobacteria ecosystems can be detected quickly and accurately. Artificially intelligent monitors will be taught to look at the latest satellite data to spot new pollutant releases from factories and ships or locate dangerous oil spills. Systems such as Iceye, Planet, and Spire are on the watch for pollution today. The systems of tomorrow, when linked to the fastest broadband systems, may be a hundred to a thousand times more capable.

Humankind is not adept in assessing risk, especially in the long term when it requires deliberative thinking. Much of human sensing systems are intuitive and immediate, and thus, reflexive. Potential food that tastes too bitter or smells bad is rejected. If one sees a forest fire raging or smells the smoke, one most likely would flee. If a storm is blowing loudly and one hears a tornado warning siren, one would most likely head for the basement.

But when risks become longer term and require mathematical, or there is a need for scientific calculation to understand, then human support for action can break down and become lethargic. Even if people are warned of the fatal dangers of smoking cigarettes or eating over-rich food, they persist. Prof Ralf Schmalzle, a biological Psychologist at Michigan State University, who studies the psychology of human risk assessment, has explained the problem in this way: "Knowledge [of risk] is not enough to motivate."

Microplastic poisoning, acidification of the seas and oceans, emission of hydrocarbons and carbon dioxide into the atmosphere, and so on are difficult to control. This is because these forms of pollution are globally pervasive, they have longer-term impact that may not be immediately noticeable, and their reform could restrain economic growth and employment.

The time has come for action. This involves continuing with the international efforts represented by the Paris Accord of 2015 and the next major initiative in Glasgow, Scotland, in November 2021. But it also needs to include lead nations more firmly committing to a more significant set of environmental reforms. These would include two key features.

Feature 1: A series of major environmental reforms aimed at significant improvements, with a timetable set using quantified objectives and year-by-year targets, similar in scope and design to the CAFÉ fuel efficiency legislation started by the United States in the 1970s. Its performance would be verified by satellite system, high-altitude platform systems, and smart monitors on the ground and in the oceans.

Feature 2: Unlike international objectives that are discretionary in scope, this would be backed by national and possibly state or regional laws and would indeed have firm enforcement provisions with the power of law enforcement behind it.

It is time to take on the four "modern horsemen" of the apocalypse. It is still not too late.

References

1. Sean, C.: A Series of Fortunate Events, pp. 24–40. Princeton Press, Princeton (2020)
2. CAFÉ Public Information Center. https://one.nhtsa.gov/cafe_pic/CAFE_PIC_fleet_LIVE.htm. Last accessed 4 Jan 2021
3. Foo, S., Asner, G.: Scaling up coral reef restoration using remote sensing technology. Front. Mar. Sci. **2013** (2019) https://www.frontiersin.org/articles/10.3389/fmars.2019.00079/full
4. Podcock, R.: New satellite reveals places on Earth most at risk from ocean acidification. Carbon Brief, February 2, 2015. https://www.carbonbrief.org/new-satellite-reveals-places-on-earth-most-at-risk-from-ocean-acid
5. Ozkara, A., Akyıl, D., Konuk, M.: Pesticides, Environmental Pollution, and Health, Environmental Health Risk – Hazardous Factors to Living Species, Marcelo L. Larramendy and Sonia Soloneski. IntechOpen (2016). https://doi.org/10.5772/63094. Available from: https://www.intechopen.com/books/environmental-health-risk-hazardous-factors-to-living-species/pesticides-environmental-pollution-and-health
6. Menon, S: Mercury Guide, March 10, 2016. https://www.nrdc.org/stories/mercury-guide#:~:text=King%20mackerel%2C%20marlin%2C%20orange%20roughy,should%20avoid%20eating%20these%20fish

7. Dunley. S: FDA, USDA combat consumer opposition to GMOs. March 9, 2020. https://www.foodbusinessnews.net/articles/15576-fda-usda-combat-consumer-opposition-to-gmos

8. Water borne parasites. https://www.cdc.gov/parasites/water.html#:~:text=Globally%2C%20contaminated%20water%20is%20a,(Crypto)%2C%20and%20giardiasis. Last accessed 15 Sept 2020

9. Fukushima Daiichi Accident. Updated May 2020. https://www.world-nuclear.org/information-library/safety-and-security/safety-of-plants/fukushima-daiichi-accident.aspx#:~:text=Following%20a%20major%20earthquake%2C%20a,in%20the%20first%20three%20days

10. Chu, J.: MIT News, July 20, 2015. Ocean acidification may cause dramatic changes to phytoplankton. https://news.mit.edu/2015/ocean-acidification-phytoplankton-0720

11. The United Nations Environmental Programme. https://www.unenvironment.org/about-un-environment/what-we-do. Last accessed 20 Sept 2020

5

Biochemical and Nuclear Weapons

If only I had known, I would have become a watch maker.
–Albert Einstein after the development of the atomic bomb

The risks that the leaders of a rogue state will use nuclear, chemical or biological weapons against us or our allies is the greatest security threat we face.
–Madeleine Albright, Former U.S. Secretary of State

Introduction

For young people, the threat posed by biochemical or nuclear weapons might feel abstract. But there are many Americans still alive who lived, at least as children, in the time of the Cold War. They remember being marched into hallways and sitting cross-legged with their hands over their necks and heads almost to the floor in a protective position against a Soviet nuclear bomb. There are still even a few survivors of the atomic bombings of Hiroshima and Nagasaki in Japan. They endured the horror of the blast followed by the devastating legacy of radioactive contamination and radiation sickness that lasted for decades after. As this chapter was being written in the fall of 2020, a ceremony was being carried out in the United States and Japan observing the 75th anniversary of this event.

Today, there are more than 13,000 nuclear warheads and bombs worldwide. These nuclear weapons exist not only in the United States and Russia, but also in China, India, Pakistan, the NATO countries, plus other locations as well. Despite the Nuclear Non-Proliferation Treaty, these other countries include North Korea, Israel, and perhaps soon Iran. There is also a ban on the use of biochemical weapons, but this has not stopped the use of such weapons in civil wars within Syria and Iraq. Nonproliferation treaties against the expansion of nations with nuclear weapons, and other treaties against the use of biochemical instruments, have to date served only to retard the use of such weapons.

J. N. Pelton, *Space Systems and Sustainability*, https://doi.org/10.1007/978-3-030-75735-9_5

Increasingly, the monitoring of efforts to create and test nuclear weaponry and to stockpile biochemical weapons have had to rely on space systems. In the future, this will include high-altitude platforms, unattended autonomous systems (UAS), and drones. It is only through a global system of electronic, optical, and other detection methods that the spread of such weapon systems can be halted or at least greatly curtailed.

This chapter explores the disarmament treaties and agreements that seek to limit the spread of biochemical and nuclear weapons, along with the use of space systems within this area, particularly with regard to the delivery and distribution of such weapons. In addition, it analyzes the role that space-based systems might use to aid disarmament and monitoring systems.

Current international mechanisms that seek to prohibit the spread of weapons of mass destruction are only as effective as the mechanisms for their detection. This includes capabilities to detect their development, manufacture, stockpiling, testing, or transfer across international borders. It must be acknowledged up front that even the most sophisticated space-based systems are better for some applications than others when it comes to arms control. For instance, nuclear weapon development, storage, and transport can perhaps most easily be tracked and monitored. The global monitoring of chemical weapon production and facilities is a greater challenge.

Biological weapons are the greatest challenge. The recent Covid-19 was not a biological weapon, but its spreading patterns might not differ all that much from an intentional biological weapons attack. It is true that similar types of pathogens could have been developed in labs and released as biological weapons. There are tens of thousands of places in countries like China, Russia, the United States, and India, and certainly thousands of places in countries such North Korea or Iran that could be used to develop biological weapons. In the case of biological agents, there is no need for large storage areas to keep harmful bacterial agents; the delivery systems could be a vial and needle or simply a person who is willing to be injected. This makes any detailed monitoring system incredibly hard. As we have seen, however, space-based systems can be used to detect the onset of virus infections.

The 75th anniversary of the World War II atomic bombings in Japan would seem to be an appropriate time to launch a worldwide effort to halt the development and use of instruments of mass destruction. Clearly, different strategies need to be invoked for each type.

The only approach to making the world sustainable for the longer term is finding a solution to not only curtail the spread and use of weapons as instruments of war but also to develop a new global perspective that transcends national, social, linguistic, cultural, and religious boundaries. Such a transformation would ultimately require what some have called a *planetary conscience* or *planetary awareness*. This has been expressed by Futurist Victor V. Motti, Director of the World Future Studies Federation, in his book *A Transformation Journey to Creative and Alternative Planetary Futures* in the following way: "When our awareness jumps to the planetary level and there confronts associated challenges, we will be better prepared for a planetary era" [1].

But such a view is not even close to today's reality. Today's world unfortunately requires protective action at all levels: international treaties, conventions, and regulatory processes to forbid the development and use of weapons of mass destruction, and sophisticated monitoring and detection systems to prevent their use around the world. The first step in considering what to do about nuclear and biochemical weapons is to understand exactly what these weapons are, where they are located, and how they might be used.

Biochemical Weapons

The United States formally discontinued its biological and chemical warfare program as of 1969, but it still has 10 officially listed sites that are test and storage facilities for biochemical agents and toxins. Most countries, however, do not divulge these locations, although satellite surveillance and other intelligence activities can be used to help identify such facilities as they exist around the world.

Most of the biochemical weapons that now exist are generally known, but where they are stockpiled and the means that might be used to distribute them are highly protected secrets. Chemical weapon agents are described in Table 5.1, as compiled by the eMedicine.com organization [2]. The U.S. Health and Human Services and U.S. Department of Agriculture also have an official list of toxins and biological agents that is provided in Table 5.2 [3]. The number of chemical weapons, toxins, poisons, and biological agents in these two tables reveals the enormous difficulty involved in conventional border protection processes, such as airport security inspection systems or drug enforcement agencies.

Table 5.1 Types of chemical weapons and crowd control agents

The various generic types of chemical weapons	
Categories of chemical weapon agents	Specific types of chemical weapon agents
Nerve agents	Sarin, soman, cyclohexylsarin, tabun, and VX
Vesicating or blistering agents	Mustard gases and Lewisite
Choking agents or lung toxicants	Chlorine, phosgene, and diphosgene
Cyanides	All types of cyanides
Incapacitating agents	Anticholinergic compounds
Lacrimating or riot control agents	Pepper gas, chloroacetophenone, and CS
Vomiting agents	Adamsite

Source: eMedicine.com

This is not a comprehensive list of chemical weapons, but it is generally indicative of the types of chemical weapons that can be produced globally. It would be possible to use more conventional poisons in a terrorist attack, for instance, to poison a water supply

Table 5.2 U.S. Health and Humans Services and U.S. Department of Agriculture Listing of Biological Agents and Deadly Toxins (2020)

U.S. Health and Human Services biological agents and toxins 7CFR Part 331, 9 CFR Part 121, and 42 CFR Part 73	
HHS-listed biological agent	**HHS-listed toxin**
Bacillus cereus biovar anthracis	Abrin toxin
Coxiella burnetii	Botulinum neurotoxins toxin
Crimean-Congo hemorrhagic fever virus	Botulinum neurotoxin-producing species of Clostridium toxin
Eastern equine encephalitis virus	Conotoxins (short, paralytic alpha conotoxins containing the following amino acid sequence X1CCX2PACGX3X4X5X6CX7) toxin
Ebola virus	Diacetoxyscirpenol toxin
Francisella tularensis	Ricin toxin
Lassa fever virus	Saxitoxin toxin
Lujo virus	
Marburg virus	
Monkeypox virus	
Reconstructed replication competent forms of the 1918 pandemic influenza virus (reconstructed 1918 influenza virus)	
Rickettsia prowazekii	
SARS-associated coronavirus (SARS-CoV)	
HHS South American-specific biological agents	**Hemorrhagic fever viruses**
Chapare	Infectious virus
Guanarito	Infectious virus
Junin	Infectious virus

(continued)

Table 5.2 (continued)

Machupo	Infectious virus
Sabia	Infectious virus
HHS South American toxins	**Toxins**
Staphylococcal enterotoxins (subtypes A, B, C, D, and E)	Toxin
T-2 toxin	Toxin
Tetrodotoxin	Toxin
Tick-borne viruses	**Encephalitis complex (Flavi) viruses**
Far Eastern subtype	Virus
Siberian subtype	Virus
Kyasanur forest disease	Virus
Omsk hemorrhagic fever	Virus
Variola major virus (smallpox virus)	Virus
Variola minor virus (alastrim)	Virus
Yersinia pestis	Virus
Overlap select agents and toxins	
Bacillus anthracis	Bacillus
Bacillus anthracis Pasteur strain	Bacillus
Brucella abortus	Bacillus
Brucella melitensis	Bacillus
Brucella suis	Bacillus
Burkholderia mallei	Bacillus
Burkholderia pseudomallei	Bacillus
Hendra virus	Virus
Nipah virus	Virus
Rift Valley fever virus	Virus
Venezuelan equine encephalitis virus	Virus
U.S. Department of Agriculture biological agents and toxins	
7CFR Part 331, 9 CFR Part 121, and 42 CFR Part 73	
Virus	**Toxin**
African horse sickness virus	*Mycoplasma capricolum*
African swine fever virus	*Mycoplasma mycoides*
Avian influenza virus	
Classical swine fever virus	
Foot-and-mouth disease virus	
Goat pox virus	
Lumpy skin disease virus	
Newcastle disease virus	
Peste des petits ruminants virus	
Rinderpest virus	
Swine vesicular disease virus	
USDA Plant Protection and Quarantine (PPQ) listing	
Coniothyrium glycines (formerly, *Phoma glycinicola* and *Pyrenochaeta glycines*)	
Ralstonia solanacearum	

(continued)

Table 5.2 (continued)

Rathayibacter toxicus
Sclerophthora rayssiae
Synchytrium endobioticum
Xanthomonas oryzae

Source: U.S. Government

Note: This was the list as of November 15, 2020. It is extremely difficult to detect the illegal transport of these agents and toxins internationally without intelligence alerts. While there are screening systems that might be used to detect guns or explosives, currently, there are no comparable methods for screening to intercept illegal transport of biochemical substances

International Legal Conventions, Protocols, and Treaties

In the aftermath of World War I, there was an attempt to control and prohibit the use of biochemical weapons. After extended negotiations at the newly established League of Nations, the Geneva Protocol was signed on June 17, 1925. The full name of the international agreement was "The Protocol for the Prohibition of the Use in War of Asphyxiating, Poisonous or other Gases, and of Bacteriological Methods of Warfare."

The protocol did not even broach the topic of possible toxin use by terrorists [4]. The many gaps in the Geneva Protocol and subsequent technological developments led to the negotiation of a new agreement that was signed in London, Washington, D.C., and Moscow on April 10, 1972. The convention as negotiated in 1972 is now agreed to by a large number of nations. It is formally known as the "Convention on the Prohibition of the Development, Production and Stockpiling of Bacteriological (Biological) and Toxin Weapons and on their Destruction." It is usually referred to as the Biological Weapons Convention (BWC) or the Biological and Toxins Weapons Convention (BTWC). The four core elements of the agreement are contained in Articles 1–4:

1. *Article 1.* Never to develop, produce, stockpile, or otherwise acquire or retain: (1) biological agents or toxins of types and in quantities that have no justification for peaceful uses; and (2) weapons, equipment, or means of delivery designed to use such agents or toxins for hostile purposes.
2. *Article 2.* To destroy or divert to peaceful purposes all agents, toxins, weapons, equipment, and means of delivery specified in Article I in their possession, or under their jurisdiction or control.

3. *Article 3.* Not to transfer or in any way to assist, encourage, or induce any entity to manufacture or otherwise acquire any of the agents, toxins, weapons, equipment, or means of delivery specified in Article I.
4. *Article 4.* To take any necessary measures to prohibit and prevent the development, production, stockpiling, acquisition, or retention of any of the agents, toxins, weapons, equipment, and means of delivery specified in Article I under its jurisdiction or control [5].

The Biological Weapons Convention is now widely supported by the overwhelming majority of the world's nations. Currently it has been ratified by 183 state parties, including Palestine. In addition, there are still four remaining signatories (Egypt, Haiti, Somalia, and Syria) whose legislatures have not yet formally ratified this Convention. As of November 2020, ten states have neither signed nor ratified the BWC: Chad, Comoros, Djibouti, Eritrea, Israel, Kiribati, Micronesia, Namibia, South Sudan, and Tuvalu [6].

One of the more interesting features of this agreement is the recognition that, as conditions change, it is necessary to have a meeting of state parties (MSP). This MSP allows a discussion of issues of concern, consideration of innovations and finance, and modifications as times and technologies change. There is an annual Meeting of Experts as well, known as (MX) [7].

Limitations on Nuclear Weapons

The Treaty on the Non-Proliferation of Nuclear Weapons (NPT) is the basic and most important instrument of international law concerning the global effort to halt the spread and limit the use of nuclear weapons. Although it has not been perfect, it has undoubtedly helped with containment.

The NPT has also served to promote international cooperation in the peaceful uses of nuclear energy under the provisions of Article IV, which states: "Nothing in this Treaty shall be interpreted as affecting the inalienable right of all the Parties to the treaty to develop research, production and use of nuclear energy for peaceful purposes without discrimination and in conformity with Articles I and II of this Treaty" [8]. Some have questioned whether encouraging the spread of nuclear reactors and nuclear power plants has been the right decision.

The NPT was formally opened for signature in 1968 and collected sufficient signatures to enter into force in 1970. On May 11, 1995, when its original 25 years had ended, an agreement was reached to extend it indefinitely. At this time, the NPT represents the only broadly binding commitment to the longer-term goal of disarmament, even supposedly by states that possess nuclear weapons. As of November 2020, a total of 191 states have joined the treaty. More countries have ratified the NPT than any other arms limitation and disarmament agreement [9].

Under the treaty, there are five official nuclear states: China, France, Russia, the United States, and the United Kingdom. These countries are also permanent members of the U.N. Security Council. Clearly, the NPT has not completely succeeded in halting countries from joining the nuclear weapons club. India, Pakistan, and North Korea are now nuclear powers and not members of this treaty. North Korea actually acceded to the NPT in 1985 and then withdrew on January 20, 2003. Israel has also not acceded to the NPT and is thought to have this capability, although its program is shrouded in secrecy. Iran acceded to the NPT at the outset and is still a member of the treaty, but has had aspirations to develop its own nuclear weapons program [10].

There have been aspirations to find a way to go beyond the NPT in order to limit the production and even reduce the amount of fissile materials that can be used to create nuclear weapons. This process started with what is called the International Panel on Fissile Material. Its responsibility is to monitor the global availability of highly enriched uranium (HEU) and plutonium. A few years ago, the Panel estimated that there was about 1340 tons of highly enriched uranium (HEU), while the amount of separated fissile plutonium was estimated to be about 520 tons. This is sufficient to create many thousands of nuclear warheads. The Bulletin of the Atomic Scientist has spent many decades seeking to further reduce the stockpiling of nuclear weapons and the production of fissile materials, but without great success.

Within the International Panel on Fissile Material, the possibility of a new treaty was consistently raised, wherein it would address not only nonproliferation but also the limitation or outright ban of the production of fissile material. This new treaty would be multilateral in scope and would need to be completely verifiable. The verifiability has been the constant sticking point. Up to now, such verification has not been possible; this is a key issue involving new space capabilities and data analytics that will be returned to shortly.

There have been many attempts within the annual Conference on Disarmament (CD) to start the negotiation of such a treaty. As far back as 1995, the Conference on Disarmament adopted a report (CD/1299) that referred to such a proposed initiative as a "Fissile Materials Cut-off Treaty (FMCT)." This effort was led by Ambassador Shannon of Canada. The report stated that the objective would be to seek "the most appropriate arrangement to negotiate a treaty banning the production of fissile material for nuclear weapons or other nuclear explosive devices".[1]

---

[1] Conference on Disarmament report (CD/1219) 1995, "Fissile Materials Cutoff Treaty."

This report called for creating an ad hoc committee to explore the possibility of a cut-off procedure and verification process stopping the global production of fissile materials. But this process was never started. There were additional efforts outside of the Conference on Disarmament to see whether some progress might be made in this direction. In 2012, the U.N. General Assembly managed to pass a resolution 67/53 that asked the Secretary General to survey the membership and report on the possible feasibility of negotiating a treaty that would ban the future production of nuclear or fissile materials such as plutonium or HEU. The resolution specified that an international panel of governmental experts on the subject should be created. The key task for this panel of experts was to address possible inspection processes that might be used to enforce such a treaty. Again, no significant progress was made.

In 2016, the U.N. General Assembly moved again, adopting resolution 71/259. Once more, this resolution specified the creation of an expert preparatory group that could explore the feasibility of a high-level fissile material cut-off treaty (FMCT). Canada and the European Union have consistently championed this idea. However, the lack of active support from Russia, China, the United States, and other nuclear weapons states has delayed progress. Most recently, during 2019 and 2020, actions by Russia, China, and the United States to develop a new array of hypersonic delivery systems for weapons, including nuclear weapons, seriously undercut efforts to move the FMCT-type initiative forward [11].

The Conference on Disarmament and the U.N. Office of Disarmament Affairs (UNODA) switched to a new line of thought known as a Treaty on the Prohibition of Nuclear Weapons (TPNW). By resolution 71/258, the General Assembly decided to convene in 2017 a United Nations conference to negotiate a new legally binding instrument to prohibit nuclear weapons, which would aspire to their total elimination. The General Assembly resolution encouraged all member states to participate in the conference, along with international organizations and civil society representatives.[2] This led to the two-part conference held in March 2017 and June 2017 at the U.N. Headquarters in New York (see Fig. 5.1). A total of 140 countries participated in some element of the negotiations. As of September 20, 2017, the Treaty on the Prohibition of Nuclear Weapons was adopted and opened for signature by states at the United Nations. There were 122 votes in favor, 1 vote against, and 1 country abstaining [12].

The result was a treaty that has now received wide support from many signatory nations. The number of countries that have signed and adhered to this

---

[2] Resolution 71/258, U.N. General Assembly, 2017.

Fig. 5.1 The 2017 conference to negotiate the Treaty on the Prohibition of Nuclear Weapons (TPNW) at the U.N. Headquarters in New York. (Courtesy of the United Nations)

new TPNW was sufficient for it to come into force as of January 22, 2021. Yet despite the growing number of countries that support the TPNW, there is a stark counter-reality: none of the countries possessing or thought to possess nuclear weapons capabilities, including China, France, Israel, India, North Korea, Pakistan, the United Kingdom, the United States, or Iran, have signed on. Indeed, just before the critical milestone of 50 countries was reached for it to come into force, the United States sought to block it. The United States tried to request that member states to the Treaty withdraw their instruments of ratification or accession. The move indicated that the United States and other supportive nuclear weapons states (NWS) believed that the treaty lacked the necessary verification procedures to succeed in practice. It contended that the agreement would not achieve the stated objectives and that it in fact "turns back the clock on verification and disarmament and is dangerous" [13].

Another relevant consideration is international agreements to limit missile systems that have the capability to strike other nations. The most pertinent is the New-START Treaty between the United States and the Russian Federation that entered into force as of February 18, 2011, with an initial duration of 10 years. This treaty set overall limits for both countries of 700 missile and bomber systems that could be equipped and deployed with nuclear devices, a total of 1550 nuclear warheads, in recognition of missiles or bombers that might be equipped with multiple devices, and a total of 800 deployed and nondeployed missile and bombers [14].

The status of this sort of strategic arms reduction treaty going forward as of the start of 2021 was very much in doubt. An extension had not been negotiated and it was set to expire as of Feb. 21, 2021. On Dec. 17, 2020, President Putin made a public statement asserting that United States' demands for new agreement provisions were stopping the extension. The Trump Administration's demand that China needed to be a part of a trilateral agreement in the summer of 2020 had gone nowhere [15]. Putin stated the lack of an agreed extension had led to the decision to develop and deploy hypersonic missile systems.

At the eleventh hour, however, President Biden assumed office. His transition team had been aware of the urgency of this matter and an extension of this vital Treaty for another 5 years was signed with the United States and Russia on February 3, 2021. In signing this Treaty extension, Secretary of State Antony Blinken stated: "Extending the New START Treaty ensures we have verifiable limits on Russian ICBMs, SLBMs, and heavy bombers until February 5, 2026" [16].

Weapons of Mass Destruction in Outer Space

Clearly, there is lack of progress for nuclear and missile arms limitations (especially in the context of hypersonic missile and advanced robotic aircraft systems) between the United States, the Russian Federation, China, and other countries. This will be one of the many challenges that the new Biden Administration must take up as a matter of urgency. The other challenge that the Biden Administration faces will be the consequences that come from the Trump Administration's creation of a "Space Force," along with the various provocative statements that President Trump made about confrontations in space. He has called for the development of space weapons and proclaimed: "Space is the world's newest war-fighting domain...Amid grave threats to our national security, American superiority in space is absolutely vital. And we're leading, but we're not leading by enough. But very shortly we'll be leading by a lot" [17].

Colleague Joan Johnson-Freeze, of the Naval War University in Rhode Island, has said: "What the Trump administration did is cross a Rubicon that every prior administration has hesitated to do, and that is to advocate overtly for development of space weapons" [17].

Biden's space diplomacy is going to be challenged in many ways. Deescalating the space-based rivalries will have to precede new efforts to use space systems as a tool for monitoring weapons systems in any new peace agreements. The effort to ban or limit nuclear weapons and biochemical weapons, thus, must be addressed in terms of not only land, sea, and missile launches but also outer space.

The Outer Space Treaty (OST) of 1967 sought to ban the deployment of weapons of mass destruction from deployment in outer space. It entered into force on October 10, 1967, and currently has 110 state parties, with another 89 countries that have signed it but have not yet completed ratification. Very significantly, this international agreement includes most spacefaring nations.

Article IV of the Outer Space Treaty states that no member states should do any of the following:

- Place in orbit around the Earth or other celestial bodies any nuclear weapons or objects carrying weapons of mass destruction (WMD)
- Install WMD on celestial bodies or station WMD in outer space in any other manner
- Establish military bases or installations, test "any type of weapons," or conduct military exercises on the moon and other celestial bodies [18]

Unfortunately, neither the term "weapons of mass destruction" nor the term "celestial body" is precisely defined. Nevertheless, most legal scholars would contend that this would include all types of chemical, biological, or nuclear weapons, and "celestial body" would certainly include the Sun, all the planets, and the moons of the planets. It also prohibits military activities on celestial bodies and establishes rules governing how space exploration and resource use should be done exclusively for peaceful purposes.

Presumably, the treaty still allows signatory nations to launch ballistic missiles, use hypersonic weapon delivery systems, or employ other such vehicles that fly in suborbital trajectories, as long as these weapons do not go into orbit. Yet, the OST does repeatedly emphasize that space is to be used for peaceful purposes. If broadly interpreted, the treaty would prohibit the launch of any type of weapons systems either into Earth orbit or into deep space. This wider interpretation would make all of outer space a nuclear and biochemical-free zone. But it also raises a practical issue related to potentially hazardous asteroids and comets, which might only be defended against by the use of a nuclear weapon [19].

The Moon Agreement of 1979, formally known as the "Agreement Governing the Activities of States on the Moon and Other Celestial Bodies," represented a further attempt to establish the peaceful use of outer space with regard to the Moon. Article 1 indicates that all of the "provisions of this agreement relating to the Moon shall apply to other celestial bodies in the Solar System" [20]. Article 3 goes on to spell out the specific meaning of peaceful uses of the Moon: there is to be no weapons system or military installation on the Moon, no weapons systems or hostile action on the Moon, or any type of weapons system in orbit or even in trajectory around the Moon.

The difficulty with the Moon Agreement, as in the case of the Treaty Prohibiting Nuclear Weapons (TPNW), is that none of the spacefaring nations of the world have signed on as signatories. Although there are well over the number of countries needed for the agreement to go into force, the absence of these powerful signatories creates practical difficulties, placing the countries that lack significant space capabilities in the position of regulating those that do have such capabilities. Opponents who aspire to exploit the resources of the Moon have often criticized the agreement as a failed attempt to create meaningful new international law [12].

New Space Technology and Applications

There is a range of capabilities that could be effectively used to track the world's nuclear weapons. The satellite-based Automatic Identification Service (AIS) is broadly applied to most ships at sea for safety and also protection against smuggling and illegal fishing. Nearly a dozen low Earth orbit satellite systems now in orbit or planned for deployment offer AIS services. These include Orbcomm, Globalstar, Iridium, and many other of the new LEO constellations.

Recently, the new Hawkeye 360 LEO constellation has been deployed. It is equipped to provide radio frequency (RF) geolocation on a global basis. It can, therefore, specify the location of all radio communications usage globally and can provide additional information about ships or vehicles operating in unauthorized areas. In one instance, an Ecuadorian patrol discovered a Chinese fishing fleet of over 250 ship operating just outside the exclusive economic zone (EEZ) surrounding the Galápagos Islands. RF analysis conducted by these satellites discovered that many of these fishing vessels had deliberately sought to conceal their AIS tracking beacons on the occasions when they had illegally entered Ecuadorian waters. The satellite monitoring, therefore, provided exact evidence of illegal fishing by these so-called "dark vessels" in the South Pacific [21].

AIS and RF geolocation could be applied to help track the location of every nuclear device in the world. This would require that they are permanently equipped with an appropriate RFID or an AIS beacon, or Internet of Things–equipped monitors. Thereafter, there could be continuous imaging data for all known nuclear storage facilities, missile silos, etc. This system would allow an uninterrupted monitoring system for the 13,000 or so such devices that exist around the world. It could also be used to track radioisotopes employed for medical purposes, or nuclear waste materials.

The key to this approach would be to create design where the AIS signal could not be removed from a nuclear device, disturbed in its location, or replaced by a counterfeit substitute signal without instant detection. If there were a storage facility with perhaps 35 nuclear devices, the AIS signal from all 35 nuclear devices would be continuously received and then an IoT uplink to the satellite network would confirm that all 35 devices remained in place. All such systems would operate on battery systems that are periodically replaced to guarantee performance against power losses. High-quality satellite imaging operating on a 24/7 basis (i.e., radar and optical sensing) would instantaneously detect disturbed devices. If there were interruptions in the AIS signals for any nuclear device, an immediate onsite inspection by International Atomic Energy Agency (IAEA) inspectors would be required to ascertain the reason for the interruption and certify the location of the disturbed device with precise verification. A similar approach might be considered for tracking the existence of chemical laboratories or suspected weapons storage facilities.

Of course, all nuclear-equipped nation states would have to agree to such a continuous monitoring system. Countries such as China, India, Pakistan, North Korea, or Iran might be difficult to persuade. But is there any effective means or space-based process to systematically verify that a nation is truly not engaged in developing these weapons? The answer at this time is probably no. Nations would have to surrender too much of their independence to allow truly systematic verification.

Space Security and Space Situational Awareness

Efforts to track objects in Earth orbit are an ongoing challenge. Currently, there are about 22,000 objects being tracked in Earth orbit that are more than 10 cm in diameter (i.e., larger than a baseball). Most of these are operational or defunct satellites, upper stages of launched rockets left in orbit, and other forms of space debris. The new S-band radar "space fence" installed in the South Pacific will be able to track over 100,000 pieces of space junk, of which about 45% (6000 tons in mass) has already accumulated in orbit.

Space debris is expected to become an increasing problem due to the large constellations of satellites being launched primarily into low Earth orbit in the next 10 years [22]. Despite the fact that operators of these new systems have announced careful plans to deorbit the satellites at the end of life, the possibility of collisions occurring at deployment, deorbit, or by accident promises an increase in orbital space debris. A recent analysis carried out by the Aerospace Corporation of the five largest new LEO constellations projected at least one space collision or more for these systems [23].

In December 2020, one of the operators of a geosynchronous satellite network, Viasat, formally petitioned the FCC to undertake an environmental review of the SpaceX Starlink, with the intent to modify its license and change its mega-LEO constellation configuration. Viasat's petition has suggested that Starlink might constitute a risk to other satellite networks [24]. This is a subject that will be addressed again in further detail in Chap. 7.

There are also increasing fears of in-orbit space weaponization. This is fueled by new initiatives such as the U.S. Government's plan to create a so-called "Space Force," India's creation of antisatellite missiles, and Russian development of hostile rendezvous and proximity operations (RPOs) that could disable or destroy another satellite. Such a capability could allow a nation to disable or attack vital defense surveillance satellites that monitor nuclear test sites or possibly GPS navigation satellites.

These space navigational satellites have many peaceful uses, but they can also be used to guide nuclear or other weapons to their targets. There is fear today that a rogue nation such as North Korea might even place a nuclear weapon in Earth orbit. Yet satellite-based systems remain a key peacekeeping and attack warning capability, monitoring for acts of war, from missile launches to movement of tanks and troops.

As put by Major Erin Salinas of the U.S. Air Force in the 20th Space Control Squadron Detachment:

> As the threats against US capabilities in the space domain continue to grow, Space Battle Management becomes much more than just catalog maintenance. It is the overall understanding of what is occurring in the domain. To ensure effective weaponeering of SSA sensors, operators need to be empowered to act through the incorporation of intelligence, flexible tasking, rapid decision making, and platform integration. Space is our ultimate high ground; we must keep watch [25].

In January 2019, the U.S. Defense Intelligence Agency published "Challenges to Space Security," a 46-page, nonclassified report that laid out in detail U.S. military concerns about space-related threats that were particularly focused on China and Russia. This report explicitly stated: "Both states [i.e. China and Russia] are developing jamming and cyberspace capabilities, directed energy weapons, on-orbit capabilities, and ground-based antisatellite missiles that can achieve a range of reversible to nonreversible effects" [26].

New space governance arrangements are needed to slow down the threat of space-based conflict. One of the institutions at the heart of this effort is the U.N. Institute for Disarmament Research (UNIDIR) based in Geneva, Switzerland. The UNIDIR organizes ongoing meetings that include the objective of what is called PAROS, or Prevention of Arms Race in Outer Space. There have been recent suggestions from Russia and China during 2020 to start discussions within a newly constituted Group of Government Experts, with a mandate to explore a Treaty on Disarmament in Space. Such an initiative, however, continues to be considered by U.S. participants as unfeasible and unable to be verified [27].

There does not seem to be a near-term threat of nuclear weapons or other instruments of mass destruction being placed in outer space. Yet weaponization of existing space objects, monitoring of nuclear tests, cyberattacks on satellites, possible antisatellite attacks, and many other concerns seem to be escalating. Many hope that a new U.S. Administration in Washington, D.C, might be able to mitigate the rising concerns of space-based conflict.

Conclusions

Nuclear weapons attacks, terrorist attacks on nuclear power plants, and natural disasters at nuclear sites remain a serious threat to the world community. The dangers of biochemical attacks are historically much less in the public eye. Nevertheless, the unfounded rumor that the COVID-19 virus might have been released from a Chinese lab that designs biological weapons has heightened global concerns about the devastation that such germ warfare could bring to an unsuspecting world. As scientist Amit Ray has said: "The power of biological weapons is ten times more than the nuclear power. Unless we act fast with an open mind, any one of them can extinct the human race" [28]. COVID-19's powerful impact on the global economy has certainly demonstrated just how powerful a biological agent could be. If a malicious group had access to a life-giving vaccine against a biological weapon they created, it could give them enormous destructive power indeed.

New treaties that address nuclear and biochemical threats with solid monitoring and verification procedures would be a logical step forward. Space-based systems linked to the Internet of Things may effectively curb and control these types of weapons. The design, nature, location, and operation would differ for each type of weapons system. Any change or disturbance in production, storage, research, or test facilities would be instantly flagged and supplemented by on-site inspection teams.

Of the different types of satellite-monitoring systems, the one for biological weapons would be the most difficult to achieve. The intent here is not to craft the design for each system, but only to suggest that such systems might be an integral part of any future inspection agreement.

References

1. Motti, V.: A Transformation Journey to Creative and Alternative Planetary Futures, p. 41. Cambridge Scholars Press, Newcastle on Tyne (2019)
2. Types of chemical weapon agents. https://www.emedicinehealth.com/chemical_warfare/article_em.htm#types_of_chemical_weapon_agents. Accessed 20 Sept 2020
3. CDC and USDA Federal Select Agent Program. https://www.selectagents.gov/sat/index.htm. Accessed 15 Nov 2020
4. https://www.un.org/disarmament/wmd/bio/1925-geneva-protocol/
5. Biological Weapons Convention. https://www.state.gov/biological-weapons-convention/. Accessed 16 Nov 2020
6. Arms Control Association: Fact sheet on the biological weapons convention and national accessions. https://www.armscontrol.org/factsheets/bwcsig
7. Reaching Critical Will: 2019 Biological weapons convention meeting of states. https://www.reachingcriticalwill.org/news/latest-news/14458-2019-biological-weapons-convention-meeting-of-states-parties#:~:text=The%202019%20Meeting%20of%20States,2019%20Meetings%20of%20States%20Parties. Accessed 20 Nov 2020
8. IAEA: Treaty on the non-proliferation of nuclear weapons. https://www.iaea.org/sites/default/files/publications/documents/infcircs/1970/infcirc140.pdf. Accessed 20 Nov 2020
9. U.N. Office of Disarmament Affairs: Treaty on the non-proliferation of nuclear weapons (NPT). https://www.un.org/disarmament/wmd/nuclear/npt/. Accessed 20 Nov 2020
10. List of parties to the treaty on the non-proliferation of nuclear weapons. https://en.wikipedia.org/wiki/List_of_parties_to_the_Treaty_on_the_Non-Proliferation_of_Nuclear_Weapons#:~:text=The%20treaty%20recognizes%20five%20states,to%20the%20treaty%20in%201992. Accessed 20 Nov 2020
11. Stone R.: 'National pride is at stake.' Russia, China, United States race to build hypersonic weapons. Science Magazine, 8 Jan 2020. https://www.sciencemag.org/news/2020/01/national-pride-stake-russia-china-united-states-race-build-hypersonic-weapons

12. U.N. adopts ban on nuclear weapons. 20 Sept 2017. https://futureoflife. org/2017/07/07/united-nations-adopts-ban-nuclear-weapons/
13. U.S. urges countries to withdraw from U.N. Treaty that would ban nuclear weapons. PBS News. 21 Oct 2020. https://www.pbs.org/newshour/ world/u-s-urges-countries-to-withdraw-from-u-n-treaty-that-would-ban-nuclear-weapons
14. The treaty between the United States of America and the Russian Federation on measures for the further reduction and limitation of strategic offensive arms also known as the New START treaty. 2020. https://www.state.gov/new-start/
15. Reif, K., Bugos, S.: U.S. modifies arms control aims with Russia. Arms Control Association. Sept 2020. https://www.armscontrol.org/act/2020-09/news/ us-modifies-arms-control-aims-russia
16. Blinken, A.J.: On the extension of the New START treaty with the Russian Federation. 3 Feb 2021. www.state.gov/on-the-extension-of-the-new-start-treaty-with-the-russian-federation/#:~:text=Today%2C%20the%20United%20 States%20took,bombers%20until%20February%205%2C%202026
17. Kennedy, M.: Trump created the space force; here's what it will actually do. NPR. org. 21 Dec 2019. https://www.npr.org/2019/12/21/790492010/ trump-created-the-space-force-heres-what-it-will-do
18. U.N. Outer Space Treaty of 1967, U.N. Office of Outer Space Affairs. https:// www.unoosa.org/oosa/en/ourwork/spacelaw/treaties/introouterspacetreaty.html. Accessed 20 Nov 2020
19. Fact sheets. The outer space treaty at a glance. https://www.armscontrol.org/fact-sheets/outerspace#:~:text=The%20treaty%20forbids%20countries%20 from,mass%20destruction%22%20in%20outer%20space.&text=The%20trea-ty's%20key%20arms%20control,weapons%20or%20objects%20carrying%20 WMD
20. Agreement governing the activities of states on the moon and other celestial bod-ies of 1979. U.N. Office of Outer Space Affairs. https://www.unoosa.org/pdf/ gares/ARES_34_68E.pdf. Accessed 20 Nov 2020
21. Chinese fishing vessels encroaches on the Galapagos Islands, https://www.he360. com/insight/chinese-fishing-fleet-encroaches-on-the-galapagos-islands/. Accessed 20 Oct 2020
22. Madi, M., Sokolova, O.: Space Debris Peril. CRC Press, Boca Raton (2021)., Chapter 1
23. Theodore, M., et al., Aerospace Corporation: Space traffic management in the new space era. Journal of Space Safety Engineering. 6(2), 80–87 (2019)
24. Foust, J.: Viasat asks FCC to perform environmental review of Starlink. SpaceNews. 28 Dec 2020. https://spacenews.com/viasat-asks-fcc-to-perform-environmental-review-of-starlink/#:~:text=WASHINGTON%20 %E2%80%94%20Viasat%20has%20petitioned%20the,in%20space%20 and%20on%20Earth.&text=Part%20of%20the%20petition%20addresses%20 orbital%20debris

25. Maj. Erin Salinas, 20th Space Control Squadron Detachment 1: Space-situational-awareness-is-space-battle-management. https://www.afspc.af.mil/News/Article-Display/Article/1523196/space-situational-awareness-is-space-battle-management/. Published 16 May 2018

26. Challenges to Security in Space. U.S. Defense Intelligence Agency. https://www.dia.mil/Portals/27/Documents/News/Military%20Power%20Publications/Space_Threat_V14_020119_sm.pdf. Jan 2019

27. SpaceWatchGL Interviews: Renata Dwan of United Nations Institute for Disarmament Research. https://spacewatch.global/2019/07/spacewatchgl-interviews-renata-dwan-of-unidir/. Accessed 23 Nov 2020

28. Ray, A.: Nuclear weapons free world peace on the earth quotes. https://www.goodreads.com/work/quotes/55576688-nuclear-weapons-free-world-peace-on-the-earth. Accessed 23 Nov 2020

6

Asteroids and Planetary Protection Systems

The dinosaurs became extinct because they didn't have a space program. And if we become extinct because we don't have a space program, it'll serve us right!

–Larry Niven, Science Fiction Writer

A hundred years earlier, a much poorer world, with far feebler resources, had squandered its wealth by suicidally attempting to destroy its enemies by nuclear weapons, mankind against itself. The effort had never been successful, but the skills acquired then had not been forgotten. Now they could be used for a far nobler purpose, and on an infinitely vaster stage. No meteorite large enough to cause catastrophe would ever again be allowed to breach the defenses of Earth. So began Project SPACEGUARD...

–Arthur C. Clarke, Rendezvous with Rama (1973)

Introduction

Science fiction writers like Larry Niven and Arthur C. Clarke have long since questioned human intelligence, suggesting that maybe it is better to use nuclear devices to defend Earth rather than destroy people in brutal acts of war. Today, instruments with more elegance and precision may be available to use in planetary defense. Indeed, NASA's Dart spacecraft, to be launched in 2021, is set to test the ability of a high-speed impact vehicle that would divert a potentially hazardous asteroid from Earth.

A large asteroid about 6 miles (roughly 10 km) in diameter hit Earth some 66 million years ago. With the force of millions of hydrogen bombs, it smashed into the Caribbean coastal waters and firmament of Mexico, creating a massive crater from which gigatons of mass was spewed into the atmosphere.

This cataclysmic event created the fifth mass extinction. It was once known as the K-T event but is now renamed the K-Pg event. Its occurrence separated the Cretaceous Period (i.e., the time of the dinosaurs and reptiles) from the

J. N. Pelton, *Space Systems and Sustainability*, https://doi.org/10.1007/978-3-030-75735-9_6

Paleogene Period, changing the course of evolution. Perhaps close to 80% of all animal and plant species perished. The world was blanketed in darkness for years. This so-called extinction event allowed mammals to replace dinosaurs over time as the dominate life forms on the planet.

Humanity is not prepared for a repeat event. With another such event some million years away, and other pressing matters such as climate change, global pollution, and nuclear weapons threatening human civilization at a much sooner date, it is hard to get people to focus on something that could come so far down the line. Yet smaller asteroids, meteors, and bolides hit the world with much greater frequency than once thought. The challenge of getting people to believe in hidden dangers recalls to mind a common story about a Massachusetts man in the 1930s, who had received in the mail his new barometer. When he opened the package, he discovered that it was defective, showing on its dial dire hurricane conditions. He promptly threw the barometer and mailing carton into the back of his car and drove to the Post Office to ship it back. By the time he got to the Post Office, his house had blown away. Sometimes the threat you cannot see can still be quite real.

This chapter outlines the nature of the threat and warning systems now in place. It addresses the difficulty of detecting asteroids and the need to track and find smaller ones that can wipe out megacities. Additionally, it shares ideas about how to defend against these risks going forward.

Historical Background

The K-Pg event triggered earthquakes and tsunamis, with waves estimated to have risen over 200 meters high. Scientist Sean B. Carroll describes the impact of the killer asteroid from an unusual perspective:

> It is the same relative size as a BB pellet to a two-story house. But the key difference is that the asteroid travels much faster—about 50,000 miles per hour—so that upon entry into the atmosphere, the fireball's impact would be powerful enough to drill a crater 120 miles wide and 25 miles deep [1, p. 21].

The impact threw molten dirt, stone, water, and ash into the atmosphere that encircled the Earth, creating a huge cloud that blocked the sun and brought death on a worldwide scale. Nowadays, large deposits of the asteroid's iridium that were thrown and scattered around the world help provide the exact dating of this event. As Dr. Carroll summed up: "Life was baked, then frozen, then starved" (Fig. 6.1) [1, p. 25].

Fig. 6.1 Astronaut image of the crater created by the K-T/K-Pg event, as seen along the Mexico Coast. (Courtesy of NASA)

About every 100 million to 150 million years, Earth has experienced extreme temperature changes that most scientists believe led to the previous four mass extinctions. Each of these mass extinction events led saw the die-off of most species on the planet. Recent research has found that the Earth has experienced many more temperature changes than once thought. These shifts are linked to changes in CO_2 levels and are associated with subsurface cold water currents in the Atlantic Ocean [1, p. 48–52].

This pattern of course was broken when a new type of extinction event occurred 66 million years ago. The earlier mass extinctions happened over decades, or even centuries and millennia as prolonged chemical and atmospheric processes forced the Earth's atmosphere and waters to heat up. The K-Pg event, however, took place over only 3 years.

Tracking Asteroids

The ultimate questions are as follows: Can scientists truly track all dangerous asteroids? Is there some viable way to defend Earth in case one is found?

Certainly, some mitigating efforts can be taken to prepare for such a strike and mobilize for recovery afterwards. But the capability of altering an asteroid's path to avert the collision altogether would be a much more important long-term goal. Perhaps over time, humanity will be able to launch a true "Spaceguard" program, as envisioned by Arthur C. Clarke in *Rendezvous with Rama*. This would be an active global system designed to divert asteroids into new orbits so as to defend Earth from cosmic threats in a systematic way [2]. For now, there are growing efforts to track the skies via ground observatories and infrared (IR) space telescopes. The modest heat of asteroids against the near absolute zero temperature of the space allows them to be more easily detected. It is thus possible to use various tools to scan the skies.

A quick explanatory note about comets might also be useful. Potentially hazardous comets are typically bigger, faster, and even more dangerous than asteroids. Also, they come from a different place in the solar system and have many different characteristics from asteroids. Different approaches and technologies are therefore needed to defend against comets. For these reasons, comets will be addressed in the following chapter.

Shifting Spending to Planetary Defense

NASA spends over $20 billion a year on its space programs on behalf of the United States. The U.S. defense agencies, including the so-called Space Force plus the various other military forces; the Central Intelligence Agency (CIA) plus other military intelligence agencies; the National Security Agency (NSA); and the National Reconnaissance Office (NRO), probably spend at least an additional $70 billion on space-related programs. And other U.S. agencies such as the National Oceanic and Atmospheric Administration (NOAA), the Commerce Department, the Interior Department, the Department of Agriculture, and the National Science Foundation probably account for another $10 billion. That totals nearly $100 billion a year or $1 trillion over a decade. This is just for the United States; globally, several hundred billion dollars are also spent over a decade. This does not take into account the commercial space sector, which is rapidly growing.

The amount of money spent by the U.S. Government on tracking and identifying asteroids and researching methods to cope with identified threats is quite small in proportion to the trillions of dollars spent on space over 10 years. Today, perhaps as much as $25 billion over a decade is spent on asteroid threats and some form of planetary defense. This is generously estimated to be 2.5% of what the United States spends on military and strategic defense. Probably, the actual numbers are far less. The expenditure on asteroid threats is thus quite low. And the U.S. government is not alone.

In short, we need to do two things in tandem. First, spend perhaps four or five times more on asteroid tracking and planetary defense. Second, create a globally consolidated effort that eliminates wasteful overlap and pools together resources.

Why do we need to do this now? Why not next year or a decade from now? We have had a number of near misses and actual asteroid strikes. There was the significant strike early in the twentieth century in Tunguska, Siberia, by a 35- to 40-meter asteroid that could have totally destroyed a megacity like New York, Paris, Tokyo, or Beijing. As luck would have it, this superbolide landed in a largely uninhabited region of Siberia. And, as recently as 2013, there was another more recent air burst of an asteroid the size of a large bus over Chelyabinsk, Russia. This unexpected explosion did significant damage to buildings and injured a number of people. If it had entered Earth's atmosphere at a different angle, it could have done far more damage.

The possibility—indeed probability—of a truly big hit goes up each year as we move toward a global population of perhaps 12 billion and as the global population moves to 80% urban. The asteroid numbers are not increasing, but our urban targets and modern infrastructure are.

The B612 Foundation and the Unsuccessful Sentinel IR Space Telescope

Astronaut Ed Lu is President of the B612 Foundation and has lived aboard the International Space Station for half a year. He says: "I have often pondered the large asteroid impact craters on both the Earth and the Moon and realized that the entire history of life on Earth has been shaped by cosmic impacts" [3].

This is an important insight. As we saw earlier in this book, the concept of Earth as Gaia—a living planet designed to sustain life—was presented by scientist and author James Lovelock. This holistic view of Earth as a biosystem has many strengths, fostering the notion that Earth is not an inanimate object to be exploited, but is rather a living organism to be nurtured [4].

It is no accident that astronaut Edward Lu, along with my friend and colleague, the Apollo 9 astronaut Rusty Schweickart have such strong feelings about potential asteroid dangers. After serving as head of the Association of Space Explorers for some time, Astronaut Schweickart went on to found the B612 Foundation. This is a not-for-profit organization dedicated to protecting planet Earth from cosmic hazards and especially asteroid strikes. Astronaut Ed Lu then took the reins as Executive Director along with Danica Remy as

President. Remy was also cofounder of Asteroid Day, now celebrated world-wide and endorsed by the United Nations. This date was set on June 30 each year, starting in 2016. It was chosen as the anniversary of the 1908 asteroid strike in Tunguska, Siberia.

Experts from the B612 Foundation and Ball Aerospace scientists joined together almost a decade ago to design a new space telescope called Sentinel. Schweickart, in a 7-minute-long video called "Cosmic Hazards," provides a very clear explanation of the key initial objectives of the Sentinel project. This video was produced by colleague Michael Potter and the author as part of the launch of *Handbook of Cosmic Hazards and Planetary Defense* (available at https://www.youtube.com/watch?v=IJDGD73aD9s). Schweickart explains: "After we get a good fix on an asteroid's orbit we can predict its pathway for a hundred years into the future" [5].

The B612 Foundation's space telescope would have used infrared sensors to detect potentially hazardous asteroids down to as small as 35 meters in size. The attempt was made to privately finance this project with some support from NASA, with Ball Aerospace serving as the aerospace partner. Ultimately, the B612 project was not able to receive the needed funding. The unconventional approach hinged on the plan to deploy the Sentinel infrared telescope in an orbit as far away from Earth as the orbit of Venus. This would allow a very broad survey area sufficiently wide to find all potentially hazardous asteroids.

Certainly they would have found far more in their search than the current search program undertaken by NASA to track asteroids that are 140 meters in diameter or larger [6]. The Sentinel IR space telescope would have given a great deal of useful information as to just how likely another asteroid-initiated mass extinction event might be. Perhaps even more importantly, the Sentinel would have generated new data on the smaller bolides that are much more numerous and, thus, much more likely to actually hit Earth. It could, for instance, have helped track the Tunguska superbolide of 1908. In short, the Sentinel IR space telescope would have provided much greater detection capabilities (Fig. 6.2).

Fig. 6.2 The now-lapsed Sentinel Infrared Space Telescope Project with its wide field of view. (Courtesy of the B612 Foundation)

NASA Efforts

NASA has in the past repurposed the Wide-range Infrared Survey Explorer (WISE) telescope. This activity was started many years after the WISE telescope was already in orbit. Thereafter, it morphed to become the so-called NEOWISE (Near-Earth Object Wide-range Infrared Survey Explorer). In this mode, the aging IR telescope was able to find a significant number of potentially hazardous asteroids (PHAs). The NEOWISE was only reconfigured to become operational in this mode very near its end of life. Yet the afterthought program was more successful than many had thought possible.

One project by NASA, similar to Sentinel, had been put forward but is now mothballed. This large IR space telescope project was to have been known as NEOCAM. Unlike the B612 telescope Sentinel, it would be limited in its ability and designed to locate asteroids greater than 140 meters in diameter. NEOCAM now seems likely to be replaced by a series of small satellites in low Earth orbit that can be deployed at lower cost. The replacement mission, known as the Near-Earth Orbit Surveillance Mission (NEOSM) and planned for deployment in 2025, is described later in the chapter. The constellation is still wedded to the Congressional mandate of finding potentially hazardous asteroids 140 meters in diameter or larger [7].

Recalibrating NASA Priorities

NASA's new infrared telescope, launched in 2020, could also have been deployed to find dangerous asteroids, but it is instead being used to find distant galaxies and exoplanets. This is a very worthy project, but still, there should always be a clear setting of priorities. Is finding more exoplanets more important than saving Earth from destruction?

Of course, ground-based observatories are also used to find and track asteroids, with special tracking emphasis on dangerous ones that could hit Earth. There are at least 30 independent observatories preforming such tracking. They range from the Andrushivka Astronomical Observatory in Oblast, Ukraine, to the Zeno Observatory in Edmond, Oklahoma in the United States. NASA, together with the Jet Propulsion Lab, created the Near-Earth Asteroid Tracking (NEAT) system in 1995. Its 40,000 near-Earth asteroid discoveries through tracking networks are shown in Fig. 6.3.

The Pan-Starrs system alone has discovered some 10,000 asteroids, and the NASA and JPL NEAT project have collectively discovered some 40,000 asteroids. Certainly then, NASA and other space agencies do spend time and resources to find and track asteroids and other cosmic threats. But ultimately,

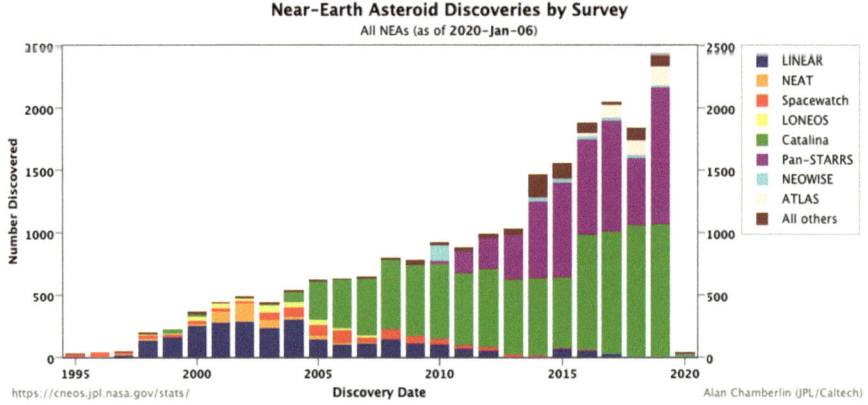

Fig. 6.3 Near-Earth asteroid tracking, led by Pan-Starrs and Catalina. (Courtesy of NASA and Alan B. Chamberlin of the Discovery Statistics Center)

their prime mission is to explore the universe, conduct space science, and help develop space applications. They do not see their mission being to protect Earth against solar storms, asteroids, and comets. After all, putting astronauts on the Moon is far sexier than looking for and cataloguing old space rocks. Surely space and planetary defense is somebody else's job, right? NASA would gladly let the Space Force, the Homeland Security Agency, or anyone else undertake that mission. Right? Unfortunately, nobody seems to know for sure.

Steps to Protect Earth from Asteroid Strikes

So, let us talk about what we know about threats from asteroids, the means we have to detect these threats, and the means we have to prevent them.

Step 1: Threat Detection

One of the dumbest sayings ever uttered is: "What you don't know can't hurt you." Rocks do fall from the sky and it is very important to know about them because they very definitely can hurt you. Just look at Fig. 6.4, which is a map compiled by NASA JPL that documents space rock events by energy level and time of day over a 20-year period.

We also think of Earth as a static mass of dirt, water, metals, and other elements with a constant mass. This too is false. Astronomers at the observatory at Arizona State University estimate that about 40,000 metric tons of so-called stardust are captured by the Earth's gravity each year. Most of that is not in large rocks but small particles that are constantly being collected. And of course this varies. During meteor showers, the amount of stardust increases.

This might seem a lot, but the amount accumulated each year is small in comparison with the Earth's 6 sextillion metric tons of mass.

The GPS satellites that help us navigate much more accurately have in some cases been equipped with special, very sensitive sensors to detect nuclear explosions and underground tests. These sensors are sensitive enough to also detect bolides (larger space rocks) and meteor hits. The data have informed our scientists that these hits are at least four times more common than we once thought [8].

Scientists use the Palermo Threat Scale to assign a threat level to all detected, potentially hazardous asteroids [9]. This is a complex system designed for scientists, whereas the Torino Impact Hazard Scale is designed for the general public. In some ways, it is much like the Richter scale for earthquakes. It

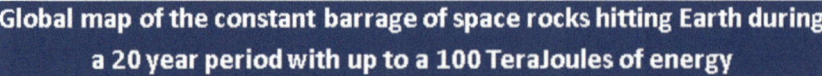

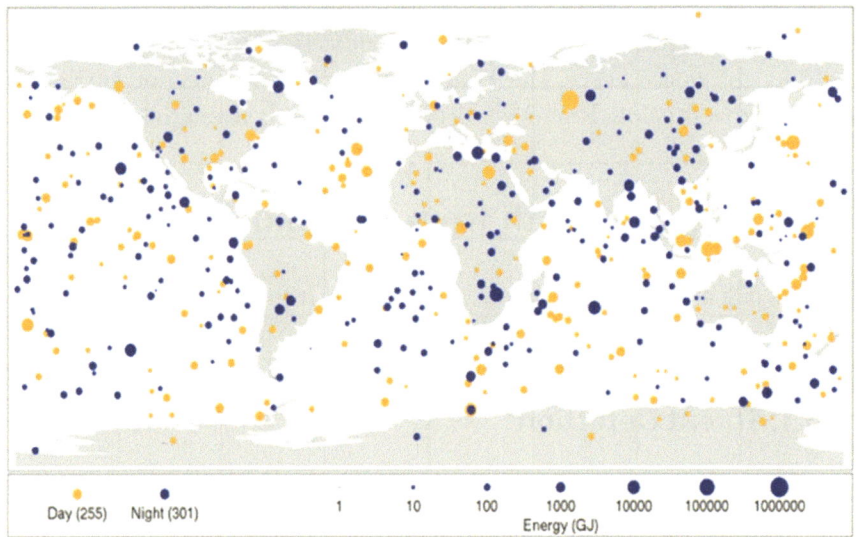

Fig. 6.4 The constant assault of bolides and smaller asteroids mapped over two decades. (Courtesy of NASA)

provides a single number from 0 to 10 to let the public know the relative danger of an impact as well as the likely consequences of different sizes of asteroids. The Torino Impact Hazard Scale is provided below in Fig. 6.5 [10].

The problem is that not all dangerous asteroids are detected until the very last minute and can enter the atmosphere or even strike undetected. This was the case of the Chelyabinsk meteor that exploded over Siberia on February 15, 2013, which we mentioned earlier in this chapter. The unexpected space rock came in at a low angle and was essentially hidden by the tremendous radiant energy of the sun. This 18- to 20-meter (55- to 66-foot) superbolide was traveling at almost 70,000 km/hour when it burst into the atmosphere at an altitude of about 30 km, or around 100,000 feet. It exploded with a great burst of energy and flash of light that was seen and felt miles away.

Although it may seem that way based on the Tunguska and Chelyabinsk incidents, Siberia has no special attractor beam—it is just that Siberia is a very large area. Figure 6.3 shows that asteroid hits are very widely distributed over land and sea.

Certainly, spacefaring countries all have programs for studying potentially hazardous asteroids and near-Earth objects. They all support the work carried

No Hazard (White Zone)	0	The likelihood of a collision is zero, or is so low as to be effectively zero. Also applies to small objects such as meteors and bodies that burn up in the atmosphere as well as infrequent meteorite falls that rarely cause damage.
Normal (Green Zone)	1	A routine discovery in which a pass near the Earth is predicted that poses no unusual level of danger. Current calculations show the chance of collision is extremely unlikely with no cause for public attention or public concern. New telescopic observations very likely will lead to re-assignment to Level 0.
Meriting Attention by Astronomers (Yellow Zone)	2	A discovery, which may become routine with expanded searches, of an object making a somewhat close but not highly unusual pass near the Earth. While meriting attention by astronomers, there is no cause for public attention or public concern as an actual collision is very unlikely. New telescopic observations very likely will lead to re-assignment to Level 0.
	3	A close encounter, meriting attention by astronomers. Current calculations give a 1% or greater chance of collision capable of localized destruction. Most likely, new telescopic observations will lead to re-assignment to Level 0. Attention by public and by public officials is merited if the encounter is less than a decade away.
	4	A close encounter, meriting attention by astronomers. Current calculations give a 1% or greater chance of collision capable of regional devastation. Most likely, new telescopic observations will lead to re-assignment to Level 0. Attention by public and by public officials is merited if the encounter is less than a decade away.
Threatening (Orange Zone)	5	A close encounter posing a serious, but still uncertain threat of regional devastation. Critical attention by astronomers is needed to determine conclusively whether or not a collision will occur. If the encounter is less than a decade away, governmental contingency planning may be warranted.
	6	A close encounter by a large object posing a serious but still uncertain threat of a global catastrophe. Critical attention by astronomers is needed to determine conclusively whether or not a collision will occur. If the encounter is less than three decades away, governmental contingency planning may be warranted.
	7	A very close encounter by a large object, which if occurring this century, poses an unprecedented but still uncertain threat of a global catastrophe. For such a threat in this century, international contingency planning is warranted, especially to determine urgently and conclusively whether or not a collision will occur.
Certain Collisions (Red Zone)	8	A collision is certain, capable of causing localized destruction for an impact over land or possibly a tsunami if close offshore. Such events occur on average between once per 50 years and once per several 1000 years.
	9	A collision is certain, capable of causing unprecedented regional devastation for a land impact or the threat of a major tsunami for an ocean impact. Such events occur on average between once per 10,000 years and once per 100,000 years.
	10	A collision is certain, capable of causing global climatic catastrophe that may threaten the future of civilization as we know it, whether impacting land or ocean. Such events occur on average once per 100,000 years, or less often.

Fig. 6.5 The Torino Impact Hazard Scale

out by the Minor Planet Center, which records all located NEOs. Likewise, they support the work of the Safeguard Institute in Italy and the newly established International Asteroid Warning Network, established under the auspices of the United Nations General Assembly in December 2015.

Yet, despite these and other worthwhile efforts, protecting Earth from space rocks ranks rather low on the priorities list. This is clearly a thankless task. You will never be in the spotlight for an event not happening—you only make the headlines if you should fail in your mission. But there is strength in numbers. A coalition of space agencies could share costs and expertise, develop improved tracking capabilities, and in time create new systems to divert asteroids from striking Earth.

Step 2: Protecting Earth

Of course, detection of near-Earth objects and tracking potentially hazardous asteroids are important undertakings. But this is just the beginning of a process. There is a lot more to be done to understand the orbital anomalies of asteroids, including gravitational or heating influences that might change them. Likewise, it is important to understand the differences in the damage associated with incoming asteroids that burst in the atmosphere, strike land, or strike the ocean. There has been important work to create computer models of such events.

Ultimately, the objective should be to develop effective ways to prevent truly dangerous asteroids from striking Earth. Certainly, measures that can create just a modest shift in orbit years in advance could be undertaken by much lower-cost missions [11]. There are many clever ideas about how such advanced defense systems could be deployed at low cost and low risk to humanity. These ideas include such projects as "Laser Bees," a mission that would use high-temperature lasers to burn potentially hazardous asteroids. The lasers would create exhausting materials from the threatening rock, driving propulsion, and moving the asteroid into a new and less threatening orbit. Another recent idea is that missions with lasers aboard could actually reshape an asteroid so that its reflectivity (or its albedo) could be changed. In this way, photons from the sun would heat up the space rock and over time be able to change its orbit [12].

One new initiative by NASA is actually undertaking a test very relevant to the active defense of Earth against asteroids. This mission is known as the Double Asteroid Redirect Test (DART). It involves the launch of the 1100-pound (500-kg) Dart spacecraft in 2021, which is to meet up in a big way with the Didymos-B double asteroid in September 2022. The plan is to crash the spacecraft into this 160-meter asteroid while traveling at a relative speed of about 25,000 miles an hour. The key is to find out if the momentum transfer is sufficient to alter the orbit of this secondary asteroid in a measurable way. The planned change in orbit is less than 1%, but this can still be sufficient to change Didymos-B's orbit around the larger Didymos-A by a few minutes. Further, the on-board camera, known as DRACO (Didymos Reconnaissance & Asteroid Camera for OpNav), is a precision imaging instrument that will reveal a good deal of information about both the Didymos A and B asteroids before the crash [13].

These approaches require quite a lot of advanced warning about a potential asteroid threat. The most cost effective and useful cosmic defense techniques depend on a more thorough inventory of potentially hazardous asteroids down to 30 meters in diameter. Ultimately, this means changing the strategic plans of the world's space agencies.

Fortunately, there are various organizations working to such ends. These include the UN Committee on the Peaceful Uses of Outer Space (COPUOS), its former Working Group on the Long Term Sustainability of Outer Space Activities, the B612 Foundation, the Association of Space Explorers, the Committee on Space Research (COSPAR), the International Science Council (ISC), the Lunar and Planetary Institute (LPI), and the American Institute of Aeronautics and Astronautics (AIAA), among others.

For a very modest fraction of what the International Space Station cost or what the Artemis program to return to the Moon will cost, space agencies could create a global project to save Earth potentially hazardous asteroids. Ultimately, there is a need to implement space shielding or asteroid orbit-diverting systems that provide a new level of global protection.

The UN General Assembly approved the proposal made by the UN Committee on the Peaceful Uses of Outer in December 2013 to create two new groups to help with potential asteroid impacts. The first of these new units was called the International Asteroid Warning Network (IAWN). The second is known as the Space Mission Planning Advisory Group (SMPAG). The official UN COPUOS description of the activities of these two units is as follows:

1. *The International Asteroid Warning Network (IAWN)* uses…communication plans and protocols to assist Governments in the analysis of possible consequences of an asteroid impact and to support the planning of mitigation responses.
2. *The Space Mission Planning Advisory Group (SMPAG)* is an inter-space agency forum that identifies technologies needed for near-Earth Object deflection, and aims to build consensus on recommendations for planetary defense measures [14].

Fig. 6.6   Asteroids pictured passing Earth. (Courtesy of ESA and Pierre Carril)

Both entities have made progress since they were established, but their work needs to be further prioritized and more resources provided. More explicitly defined monitoring and tracking efforts to identify potentially hazardous asteroids (PHAs) of smaller dimensions down to 30 meters in diameter must be undertaken. Unfortunately, there are many tens of thousands more PHAs yet to be identified (Fig. 6.6).

The focus in this chapter is on asteroids, but as will be discussed in the next chapter, comets present an even larger technical challenge in terms of effective diversion capabilities. In addition to many technical challenges, there are legal and regulatory issues to face. Many of the techniques that might allow the diversion of asteroids could also be considered space weapons. The SMPAG mechanism might address international legal instruments or collaborative processes that could be applied to joint international asteroid diversion methods. Such new capabilities would be subject to an international activation process or blocking mechanism that would prevent unilateral use by a single nation.

Such a new capability could not only provide much better protection against a cataclysmic asteroid strike, but it could also create a very powerful precedent. The fact that many astronauts and cosmonauts joined together to support the United Nations General Assembly's actions back in December 2013 is indicative of what the path forward should be. The world should listen to the judgment of the explorers who have traveled into space and seen Earth as it truly is—a quite small planet in a vast universe.

As noted earlier, the B612 was forced to abandon its Sentinel spacecraft project for tracking asteroid threats because it could not raise the necessary funds of close to a third of a billion dollars. The Planetary Society, which backed the B612 initiative, is currently supporting the most recent initiative by NASA, known as the Near-Earth Observation Mission (NEOSM). This

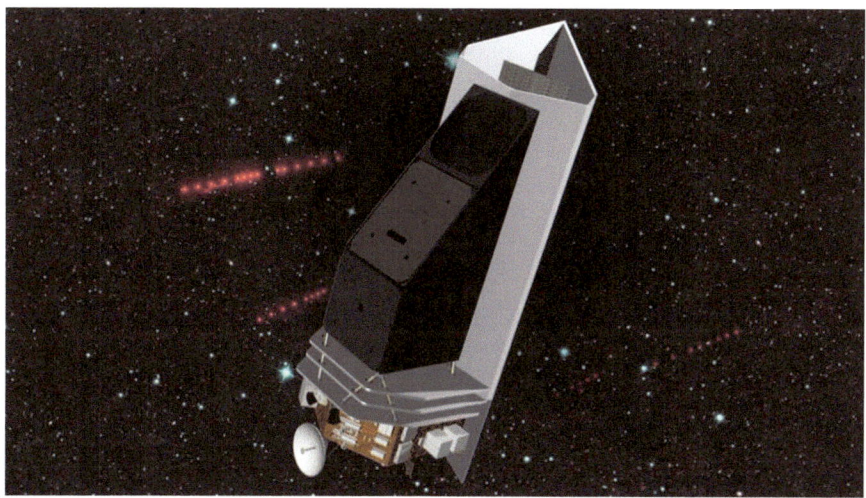

Fig. 6.7 One of the intended NEOSM small satellites to be deployed by NASA in 2025. (Courtesy of NASA)

new program would be carried out by a series of small satellites using infrared telescope lens with surprisingly large apertures on them. The initiative would replace the so-called NEOCAM that would have involved a larger infrared space telescope (see Fig. 6.7).

The current NASA plan would deploy lower-cost satellites in lower orbits. These are now being planned for possible deployment in 2025. They will be able to track potentially hazardous asteroids, but again, they are being designed to track down only up to the 140-meter or above size, as mandated by Congress. In U.S. Congress hearings held many decades ago, strategies for coping with an incoming asteroid were discussed. As a result of the first hearing, NASA was given the mandate to find and track the orbits of all potentially hazardous asteroids that were 1 km in diameter or larger. In subsequent hearings and markup, a new bill known as the George Brown NEO Survey Act was passed in 2005. The U.S. Congress at that time agreed to set a target of finding PHAs that were 140 meters or larger in size [15].

New design and deployment of smaller spacecraft and a cooperative venture among the world's space agencies could achieve the survey target for PHAs at the much more aggressive measure suggested here—that is, either 30 or 35 meters and above. Such an effort would be much more likely to protect the world's growing number of megacities that each year become more vulnerable [16]. Perhaps four or five smaller satellites might be funded by NASA, ESA, and JAXA, and perhaps the other two by China, Russia, or India. An aperture of 1 meter or so might be able to track and inventory PHAs down to a much smaller size.

Conclusions

In short, there is an urgent need to reshuffle the deck. It is hoped that the arguments presented in this book are ultimately convincing enough to spawn investment and reallocation of resources into planetary defense. Current efforts to protect our planet against such risks are too piecemeal, too under-funded, and much too inadequate when a longer-term view is taken. Space agencies need to work closely together on a global plan for detecting poten-tially harmful asteroids, centaurs, and comets, and protecting against them. This is perhaps even more urgent in terms of protecting against dangerous solar storms, space weather, and coronal mass ejections, as well as the rapidly increasing problem of orbital space debris, all of which will be investigated in the following chapters.

References

1. Sean, C.: A Series of Fortunate Events. Princeton Press, Princeton (2020)
2. Project Spaceguard. http://scifiquotes.net/quotes/146_Project-Spaceguard. Accessed 20 Oct 2020
3. Astronaut Edward Lu, Foreword to *Handbook of Cosmic Hazards and Planetary Defense*. Springer Press, Cham(2017)
4. How James Lovelock introduced Gaia to an unsuspecting world. The Guardian, 27 Aug 2010. https://www.theguardian.com/science/2010/aug/27/james-lovelock-gaia
5. Interview with Astronaut Rusty Schweickart in 2013. https://www.youtube.com/watch?v=IJDGD73aD9s
6. Lu, E.T., Reitsema, H., Troeltzsch, J., Hubbard, S.: The B612 Foundation Sentinel Space Telescope. https://b612foundation.org/wp-content/uploads/2016/05/B612-Foundation-Sentinel-Space-Telescope.pdf. Accessed 20 Oct 2020
7. Foust, J.: NASA mission to track near Earth objects takes shape. Space News. 26 Dec 2019. https://spacenews.com/nasa-mission-to-track-near-earth-objects-takes-shape/
8. Blast sensors detect more asteroid strikes than expected. NBCNews.com. 4 Apr 2014.https://www.nbcnews.com/science/space/blast-sensors-detect-more-asteroid-strikes-expected-n72486

9. Palermo technical impact hazard scale. JPL Center for Near Earth Objects. https://cneos.jpl.nasa.gov/sentry/palermo_scale.html. Accessed 20 Oct 2020

10. Torino Impact Scale, JPL Center for Near Earth Objects, https://cneos.jpl.nasa.gov/sentry/torino_scale.html. Accessed 20 Oct 2020

11. B612 Foundation. https://b612foundation.org/wp-content/uploads/2016/05/B612-Foundation-Sentinel-Space-Telescope.pdf. Accessed 20 Jan 2021

12. Ehresmann, M.: Asteroid control through surface restructuring. Acta Astronautica. Sept 2020. https://scholar.google.com/scholar?q=Asteroid+control+through+surface+restructuring&hl=en&as_sdt=0&as_vis=1&oi=scholart

13. Dart Mission. https://www.nasa.gov/planetarydefense/dart. Accessed Jan 10 2021

14. International Asteroid Day and the IAWN and SMPAG. https://www.un.org/en/observances/asteroid-day

15. H.R. 1022 — 109th Congress: George E. Brown, Jr. Near-Earth Object Survey Act. www.GovTrack.us. 2005. 16 Oct 2020. https://www.govtrack.us/congress/bills/109/hr102

16. NEOSM at a Glance. Planetary Society. https://www.planetary.org/space-missions/neosm

7

Comets and Lesser Known Cosmic Threats

Comets are like cats; they have tails and they do precisely what they want.
                                                                    –David Levy

There are more papers about dung beetle reproduction than human extinction. We might have our priorities slightly wrong.
                                                                    –Anders Sandberg

It really is a shame that through our sad neglect of wonders, hopefulness, and trust we allowed so much clutter and debris to build up in space.
                                                                    –Michael Chabon

Introduction

Somebody walking down the street does not spend much time worrying about comets smashing into Earth or wondering what a so-called "centaur" might be. This by the way is not a hybrid man-horse in the world of astronomy, but actually a comet that has been captured by the gravitational field of one of the giant planets. Thus, it has become an icy object in an unstable orbit. If it should somehow break loose from its captor's gravitation field, it could become a potentially hazardous object that could threaten Earth.

Neither do Joe Schmo nor Jane Doe worry much about the effects of a supernova on the ozone layer, or the perils of orbital debris. They have more pressing worries about paying rent on time and perhaps what movie to see on the weekend. The typical person who lives in a large city might know in the back of their mind that they depend on vital infrastructure in their daily lives. These include the systems that supply their water, operate their town's sewage system, supply their electricity and natural gas, feed gasoline to filling stations, and more. Yet for the most part, their day-to-day lives rely on one main assumption—that somebody, somehow, in their city, county, state, or country is worrying about such things for them.

The people who live on Earth are fortunate. The world is truly a "Goldilocks" place to live. It is not too big that gravity crushes human life forms. It is not

J. N. Pelton, *Space Systems and Sustainability*, https://doi.org/10.1007/978-3-030-75735-9_7

too hot or cold to survive. Spaceship Earth's atmosphere, ozone layer, geomagnetosphere, and its fortunate location in the solar system are natural protections against solar storms, asteroids, comets, supernova radiation, and other things that go bump in the night. No other planet in the solar system has so many kinds of natural protections. And Earthlike planets beyond the solar system—exoplanets—are indeed a rarity in the universe, although thousands have now been detected.

This chapter continues exploring cosmic hazards, focusing on comets, centaurs, and orbital space debris. The quite different dangers posed by powerful solar storms will be addressed in Chap. 8. To many, this may seem like science fiction, but the potential ramifications are staggering.

The Increasing Risk of Cosmic Hazards

Most global planetary defense efforts have been directed toward asteroid detection and some possible defensive actions, but there is much more to be done to truly protect the Earth and modern society.

As already noted in this book, a number of cosmic problems that threaten modern infrastructure and society are becoming more likely. This is not because of any particular rising risk out there, besides orbital space debris. For the most part, the changes that increase human vulnerability to cosmic hazards, as well as terrorist attacks, pandemics, natural disasters, and other risks, are happening down here on Earth. This topic has been mentioned quite a few times so far. Now, let us break down why this is and what it really means.

The two main changes and vulnerabilities are as follows.

Infrastructure dependence Much of the world now relies on automation and associated infrastructure for their jobs, supply chains, food, and other essentials. Here is a look at some of the most important ways the modern world has changed:

- Heavy dependence on transportation and supply chain systems.
- Heavy dependence on telecommunications, data networks, IT systems, the Internet, and automated controls.
- Heavy dependence on electrical power grids, oil and gasoline pipelines, and other sources of power.
- Automated water purification and sewage systems are essential to public health.

- Modern medical, health, police, fire, and EMT systems are also essential to public health.

Just imagine life in a penthouse condo in some of the world's tallest buildings without elevators, electricity, or firefighting equipment. All of this infrastructure that is now so pervasive around the world can be suddenly taken out by a massive solar storm or an asteroid or comet strike. This is true not only for the United States, Europe, Japan, and the wealthiest countries with the highest per capita incomes but also for India, Brazil, Argentina, China, Russia, Ukraine, Egypt, Indonesia, South Africa, Saudi Arabia, etc.

Endangered population target areas Rapid population growth and urbanization cover more and more of the world's surface. Everywhere, there is increasing urbanization. These habitation patterns all serve to increase the area where lethal impact from asteroids and large bolides can occur. The constant spread of human habitation, and urbanization in particular, also increases the risk associated with pandemics, natural disasters, terrorist or military attacks, especially with weapons of mass destruction, and so on. Natural and cosmic threats that might have affected hundreds or even thousands of people centuries ago could now be dangerous to potentially millions or even billions of people. If this danger does not come as a result of initial impact, then it could come through collateral impact. This might be because of starvation or thirst due to a lack of access to food and water. Or it might be due the loss of jobs because there is no power or communications and data networks that work. Or it might be one of dozens of other vulnerabilities that no one thought were there.

The Unique Risk of Comets

A comet impact is less likely than an asteroid strike but still is quite plausible over the long term. And as impact expert Dr. Mark Boslough of the Sandia National Laboratories in New Mexico has said, a comet strike would produce "…a much bigger explosion, a much bigger crater, much more v" [1]. He explained that asteroids travel in more predictable orbits closer to Earth and, thus, tend to strike in what can often be just a glancing blow. In contrast, comets travel in what are considered to be more dangerous orbits and at much faster velocities. They arrive from the Oort cloud at the extreme reaches of the solar system and achieve enormous speed by the time they reach the neighborhood of the inner planets. This means that the hit would more likely be a

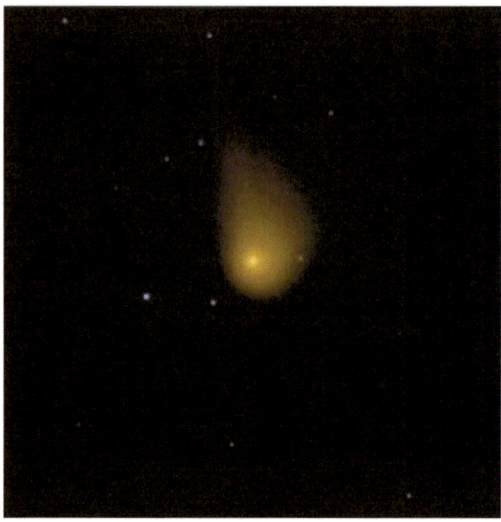

Fig. 7.1 NEOWISE IR space telescope image of the Christensen comet. (Courtesy of NASA)

head-on collision instead of a sideswipe. In terms of basic physics, a comet strike would therefore represent a much greater kinetic energy transfer.

These icy planetesimals have long been considered foreboding omens in history; this reputation could one day become well deserved (Fig. 7.1).

As David Levy, the codiscoverer of the Levy-Shoemaker comet, has said rather whimsically: "Comets are like cats; they have tails and they do precisely what they want."

The odds are very small, but the risk could be quite large. And the truly bad news is that if a comet were detected on a trajectory to hit Earth, we currently have no good defenses in place. The steep angle of attack of the comet's orbit and its much greater size and speed mean that diverting the comet would be much more difficult than diverting an asteroid. A threating asteroid would likely be smaller, less massive, slower, and on a much less dangerous attack angle. Techniques to divert asteroids such as those discussed in the previous chapter, including laser bees and changing albedos, would be ineffective against a comet, as the comet requires much more energy to change its orbit.

The only good news is that comets by population count are about 100 times less common than asteroids, and so the chances of one striking Earth are indeed much less. Further, the huge gravity wells represented by the Sun, Jupiter, Saturn, and even Neptune serve as a sort of Earth protection system. These three massive bodies offer a good chance of capturing a comet before it could hit Earth.

Fig. 7.2 Saturn's "moon" Phoebe may not actually be a moon, but rather a captured centaur comet. (Courtesy of NASA)

Most comets originate in the Kuiper Belt, but some escape and then become trapped or diverted into very irregular orbits governed by one or more of the larger bodies listed above. When this happens, the comet is designated a *centaur*. Such comets are also variously described as transitory objects and sometimes as Kuiper Belt Objects (KBOs). It is as if there were a sort of gigantic cosmic Skee-Ball game going on; it is far more likely that the comet (i.e., the Skee-Ball) will be swallowed up by these larger and more massive bodies. The Earth's target area in both size and gravitational attraction is a much harder target to hit. Captured by these planets, the comets (now known as centaurs) are still often in unstable orbits and can jump back out of the hole and again become a hazard. At the opposite extreme, they might even leave the solar system. These are typically big objects and can be much larger than a mile in size. Their velocity might be slower than an average comet, but their mass makes them dangerous and their irregular orbit makes them hard to track [2].

Many astronomers think the Saturn's moon Phoebe is in fact a centaur due to its shape and composition (Fig. 7.2).

Mitigating the Threat

As discussed in the previous chapter, the United Nations has agreed to create an International Asteroid Warning Network (IAWN) and Space Mission Planning Advisory Group (SMPAG) to defend against asteroids. But the testimony before the U.S. Congress by NASA on planetary threats from the sky focused almost exclusively on asteroids, and there is no explicit action taken to defend against comet strikes.

There is now a good range of options that could be used to divert an asteroid from hitting Earth, particularly if it is detected in reasonably good time. However, as we have seen, when it comes to comets, the answer is very dependent on its size. As Dr. David Morrison of NASA Ames Research Center explains, "If the comet is 10 kilometers across or larger (that is, if the impact carries an energy of more than about 100 million megatons), the resulting global environmental damage will be so extensive that it will lead to a mass extinction, in which most life forms die" [3].

Astrophysicist Gerrit L. Verschuur of Rhodes College in Memphis, Tennessee, who specializes in the study of comets and the amount of kinetic energy they carry, is even less optimistic. He has estimated that a comet only 1 kilometer in diameter, which would presumably be a thousand times less energetic than a 10-kilometer-diameter comet, would still be sufficient to create a mass extinction event due to its enormous velocity. (Note: The formula for the volume of a sphere is $4/3\pi r^3$. This means that a difference in volume and presumed amount of mass between a comet of 1 km and 10 km in diameter is thus 10^3 and not just a difference of 10.) [3].

In short, if a comet of a size anywhere between 1 kilometer and 10 kilometers in diameter were to be detected today on a trajectory that would impact Earth, the effective planetary defense plan would be not so much.

The greatest amount of empirical knowledge about a large comet striking a planet occurred when the many fragments of the Shoemaker-Levy comet blasted into Jupiter. All who witnessed these events via telescope or satellite-relayed images were shaken by the devastation the impacts had on the surface, creating scars the size of the entire Earth.[1]

The most promising strategy to divert a comet is a directed energy system that uses very high-powered lasers. As Professor Philip Lubin, Professor of Physics at the University of California, explains: "Directed energy in the form of photons plays an increasingly important role in everyday life…Recent advances in laser photonics now allow very large-scale modular and scalable systems that are suitable for planetary defense" [4, p. 941]. Professor Lubin has developed what he calls the "DE-STAR" system that with a sufficiently powerful stream of photons could direct energy beams into space, heating over time a potentially hazardous cosmic body to 3000 degrees Kelvin (2727

---

[1] Interview with Dr. Firooz Allahdadi, who served as lead investigator for the U.S. Air Force, 2019.

degrees Celsius/4940 degrees Fahrenheit). Lubin suggests that this is a sufficiently high temperature to vaporize most cosmic bodies [4, p. 942].

The key to creating such a capability involves at least two challenges. One challenge would be to create such a tremendously powerful laser beam. The other would be to operate what might be considered a "space weapon," which some would consider to be a violation of the Outer Space Treaty and other international agreements discussed in Chap. 5.

In terms of the first challenge, there is now underway a highly relevant initiative funded by venture capitalist Yuri Milner. Its executive director is former NASA director Dr. Pete Worden. This project, known as Breakthrough Starshot, seeks to create and connect a series of parallel power sources to form a tremendously strong laser beam system. This would propel a fleet of a thousand light sail space probes to perhaps 15–20% of light speed. Each probe would have a mass measured in only grams and dimensions measured in centimeters. Each could be powered such that in 20–30 years they would theoretically be able to travel to the Alpha Centauri star system and send video images back to Earth. Milner has along with Worden and the advisory board's advice suggested that the first mission might be ready for launch as soon as 2036. The initiative would perhaps cost between 5 and 10 billion dollars. The largest challenge of this initiative might well be to create a scaled, connected power source for the laser beam [5] (See Fig. 7.3).

Fig. 7.3 Laser array concept designed to transmit up to 100 gigawatts of power in order to launch a small starcraft to Alpha Centauri. (Courtesy of Breakthrough Starshot)

If the power for the laser is sufficiently high, it might become the prototype for a directed beam system that could be used to vaporize threatening asteroids or even a potentially hazardous comet detected in time. At the time of writing, this might be the most promising technology for a comet defense system currently in development. In addition, the regulatory processes for launching such a mission might also create international agreement for the operation of such systems for global scientific purposes and perhaps ultimately for planetary defense.

Another step should be to make sure that valuable resources are protected. This might involve such aspects as subterranean seed banks, stored medical supplies, and essential life-sustaining elements that could survive a massive hit by a comet. These protective vaults need to include essential knowledge of science, biology, medicine, engineering, architecture, mathematics, history, and key aspects of global art, law, and philosophy. In short, this would be the acquired knowledge of human society needed to rebuild a devastated human civilization. It is sometimes surprising to find what types of protective action have already been accomplished in this regard. In the U.K., the "Doomsday" project records its architectural history and also includes much of contemporary thought and history. There are protected seed banks located around the world; even Nabisco has buried in a protected locale the recipe for Oreo cookies.

Orbital Space Debris

The parallels between the increasing levels of global pollution on Earth and space debris around it are hard to miss [6]. The orbital debris problem is becoming increasingly severe in the 2020s. There are several dozen new commercial efforts to launch constellations of several thousand small satellites into low Earth orbit. Worst of all, many of them are planned to be launched into saturated orbital areas between 800 km and 1200 km altitude, where other constellations are already established. These becomes a major risk factor to operators of new and existing networks, as well as to anybody launching payloads into higher orbits or into deep space.

Currently, if proposed networks can be coordinated under existing International Telecommunication Union (ITU) procedures to minimize interference and are operated in authorized frequency bands, there are no international prohibitions against their launch into these congested areas.

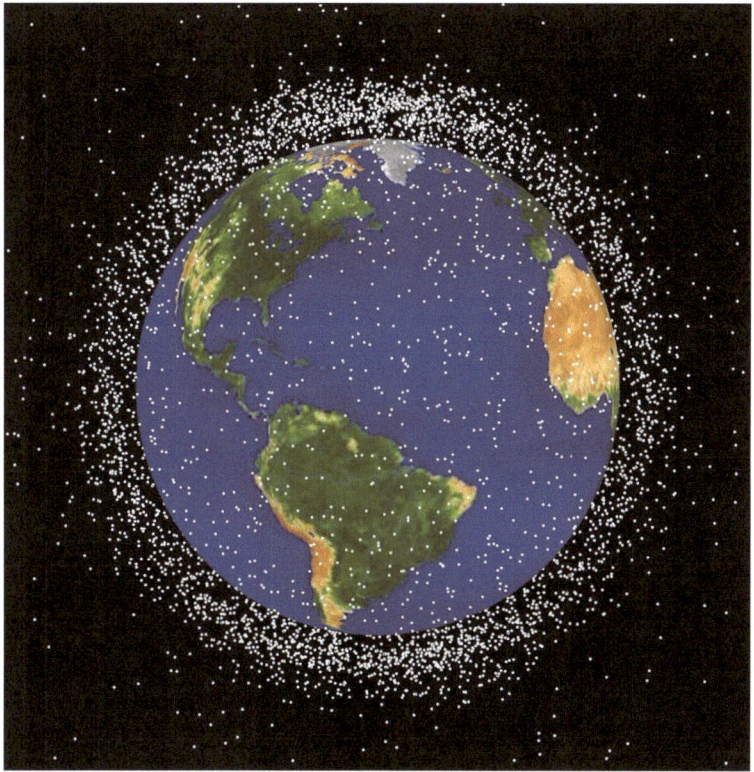

Fig. 7.4 A representation of the significant concentration of space debris in low Earth orbit. (Courtesy of NASA)

Most of them are for various forms of telecommunications and data network or remote sensing services. New applications are also arising, such as radio frequency geolocation and security applications such as automatic identification systems (AIS), which we have seen in previous chapters. While these have their benefits, the end result is that more and more operational and defunct satellites, upper-stage rockets, fairings used to cover satellites during launches, and other debris are clogging up low Earth orbit (See Fig. 7.4).

Questions to Ask

There are several key questions about the current space debris environment in Earth orbit. Let us explore them and their answers.

Are the Current U.N. Committee on the Peaceful Uses of Outer Space (COPUOS) Guidelines for Orbital Debris Removal Sufficient to Prevent Runaway Buildup of Debris (Known as the So-Called Kessler Syndrome)?

The fact that these removal guidelines are voluntary and there are no enforcement or policing powers to stop excessive deployment of new satellite constellations should be sufficient to answer this question. In short, the answer is no.

The U.N. Guidelines are inadequate for the task. The U.N. processes within COPUOS require unanimous agreement. This is a process that is generally too slow to contend with a problem that is becoming more urgent. Operators of new satellite constellations, out of their own self-interest, are themselves beginning to struggle to find responsible ways to remove defunct satellites and deploy new systems. The satellite operator ViaSat, for instance, has petitioned the U.S. FCC to conduct a formal environmental review of the LEO orbital configuration for the giant SpaceX Starlink constellation.

If U.N. regulatory processes cannot find a way beyond unanimous agreement, they will remain insufficient for many key issues, whether with regard to global pollution, climate change, or orbital debris. A global sustainability treaty could be a key first step, but it will need some teeth for it to be effective. Orbital debris control and mitigation should be a part of its scope [7, Section 2.16.4, p. 52].

Is the So-Called 25-Year Rule to Remove Satellites from Orbit After the End of Life a Helpful Guideline?

Perhaps at one time when space was not so "congested and contested," this guideline might have made sense. The average lifetime of satellites in low Earth orbit (LEO) today is around 5–7 years. Most operators of large-scale LEO constellations intend to remove their satellites quickly after end of life because they also intend to replace them with newly deployed satellites to reconstitute the constellation. The 25-year guideline is thus archaic and irrelevant in the new age of mega-LEO constellations, with thousands of satellites planned for some of these networks.

The Aerospace Corporation examined the deployment, end-of-life deorbit, and operational planning for the five largest constellations now filed with the ITU. It found that current processes for deployment and active satellite removal will be inadequate to stop rapid debris buildup. This study projected that over a 10-year period, all of these constellations would lead to new satellite collisions either during deployment or deorbit operations, or in some cases, both [8].

How Serious Is the Problem of Satellite Accidents or Collisions Today and What Is the Forecast for Coming Years?

There have already been two major satellite accidents in space, each of which created thousands of trackable debris elements. These collision events alone increased the total amount of major debris elements in LEO orbit by over 25%, or an increase of 5000 elements of significant debris (i.e., over 10 cm in diameter).

These collision events were the Chinese missile target strike on one of their old Feng Yun weather satellites in 2007, and the collision of the Russian Cosmos and U.S. Iridium commercial satellite in 2009. For the calendar year 2019, there were 63 reported violations of debris or satellite trajectories that came nearer than 4 km to the ISS and its specified safety zone. The latest forecast by the European Space Agency for new collisions occurring in Earth orbit was one every 5 years, and this was without the launch of the new mega-LEO constellations [7, p. 48].

Will It Affect Our Daily Lives?

The public at one time was concerned about a satellite, space station, or rocket falling out of the skies and hitting their house or killing someone. Space safety measures for controlled deorbit and space traffic management should reduce such risks. Launch safety is well controlled. The Space Shuttle Columbia accident that created a blizzard of debris from California to Texas was calculated to have had a 1% chance of causing a fatal airline catastrophe.

The short YouTube video "A Day Without Satellites" presents an overview of all the types of satellites services on which people depend. Satellites are today needed for communications, data networking, Internet operations, remote sensing for scores of industrial and governmental functions, weather prediction and storm tracking, climate change and pollution monitoring, space navigation and time synchronization, geodetics, global mapping, military defense and surveillance, as well as scientific purposes.

The *Kessler Syndrome* occurs when one orbital collision leads to a series of other collisions and runaway space debris buildup, endangering these systems.

In this horrible scenario, Earth's orbit would now be blanketed by millions of pieces of debris. This would prohibit future safe access to space for all types of satellite operators and potentially for all types of orbits.

The Deadliest Orbital Space Debris

The top debris concerns, as formally ranked by a panel of scientific experts, are primarily the large last stages of big rocket boosters left in orbit at around 800–1000 km altitude (See Table 7.1). Also there are many huge dead satellites such as Envisat that, if hit by another piece of debris, would create thousands of new debris elements [9].

Most new space waste comes from collisions with debris already in orbit. Removal of space debris by various means and even repurposing of debris to create new space systems are now key objectives for the future.

Other Cosmic Dangers

For years, it has been assumed that far distant supernova events would have no impact on life on Earth, but recent research has called such assumptions into question. Around 359 million years ago, there was the so-called Hangenberg event, which lasted thousands of years, spawning a major depletion of plant and even some bacterial life. Some have attributed this Hangenberg event to such factors as a significant temperature change on Earth.

But astrophysicist Brian Fields of the University of Illinois-Urbana has suggested another theory. Field research has led him to speculate that a massive supernova many light years away from Earth might have created a blast of radiation that greatly damaged the Earth's ozone layer, exposing the Earth's vegetation to a massive amount of ongoing solar radiation that included not only ultraviolet light but also highly energetic X-rays. This is only a theory, but some in the scientific community have taken it seriously, as temperature fluctuations or other explanations would presumably mean that the Hangenberg event, which divides Devonian and the Carboniferous period, should be much shorter, perhaps lasting only a decade or so and not thousands of years.

This theory can at this point only remain a theory. Yet we know that very powerful supernovae occur. The Cassiopeia A supernova, for instance, occurred some 11,000 years ago, yet its highly energetic wave of supercharged ions reached Earth just 300 years ago. Fortunately, the radiation was not

Table 7.1 The 30 most dangerous space debris elements

Rated threat level	Nation or responsible agency	Type of debris	Apogee & perigee in KM	Inclination in degrees	Date of launch
1	Russia/CIS	SL-16 Rocket Booster	848/837	71.0	3/26/1993
2	" "	" "	848/827	71.0	11/17/1992
3	" "	" "	846/843	71.0	6/29/2007
4	" "	" "	854/827	71.0	2/03/2000
5	" "	" "	844/833	71.0	10/22/1985
6	" "	" "	853/834	71.0	5/22/1990
7	" "	" "	1006/986	99.5	12/10/2001
8	" "	" "	852/831	71.0	10/31/1995
9	" "	" "	844/835	71.0	7/28/1998
10	" "	" "	845/838	71.0	11/24/1994
11	" "	" "	846/823	71.0	5/13/1987
12	" "	" "	845/841	71.0	4/23/1994
13	" "	" "	844/840	71.0	12/25/1992
14	" "	" "	850/823	71.0	9/16/1993
15	" "	" "	848/831	71.0	11/23/1988
16	" "	" "	863/839	70.8	9/04/1996
17	" "	" "	848/842	71.0	6/10/2004
18	" "	" "	841/831	7.10	3/18/1987
19	" "	" "	842/814	71.0	5/15/1988
20	" "	" "	813/801	98.6	7/10/1988
21	ESA	Envisat Spacecraft	766/764	98.1	3/01/2002
22	Russia/CIS	METEOR 3M	1013/994	99.6	12/10/2001
23	Japan	ADEOS	794/793	98.9	8/17/1996
24	Japan	H2A Rocket Booster	836/734	98.2	12/14/2002
25	Russia/CIS	SL-12 Rocket Booster	847/838	71.0	9/28/1984
26	Peoples rep. of China	CZ-2D Rocket Booster	846/791	98.7	11/20/2011
27	Russia/CIS	SL-8 Rocket Booster	995/966	82.9	3/15/1978
28	Japan	H-2 Rocket Booster	1306/860	98.7	8/17/1996
29	Russia/CIS	Cosmos 2322	854/842	71.0	10/31/1995
30	Russia/CIS	SL-8 Rocket Booster	992/861	82.9	2/05/1991

The data for this chart were derived from a paper presented at International Astronautical Congress, a cyberevent hosted by DLR of Germany. See Darren McKnight et al. "The Fifty Most Dangerous Space Debris Objects" Sept., 2020 https://iafastro.directory/iac/paper/id/55502/abstract-pdf/IAC-20,A6,2,1,x55502.brief.pdf?2020-09-10.09:43:39

sufficient to wipe out the ozone layer or greatly alter the protective shielding of the Van Allen belts.

There are some scientists who feel that serious thought should be given to artificially regenerating ozone to protect Earth. This could better help protect us from supernovae, along with other human activity such as high-altitude aircraft and plans for hypersonic craft that would fly into the highest reaches of the stratosphere. The existing, depleted ozone hole has apparently exposed animals such as frogs in the far southern latitudes to genetic mutation, and people in the southern areas of Australia to elevated levels of skin cancer.

Comets and supernovae are not the only possible cosmic concerns. Human understanding of the structure, physics, and dynamic nature of the universe remains quite elementary. True knowledge of the process that leads to the formation of stars, solar systems, dark energy, dark matter, black holes, anti-matter, and other elements of the universe is far from perfect. Daily life on Earth largely proceeds on the working premise that the world is an independent body unto itself, rather than a 6.6-sextillion-ton spaceship hurtling around the sun at 107,208 kilometers per hour.

Earth is a heavenly body, currently with close to 8 billion human passengers plus quadrillions of animal and plant organisms aboard. Earth seems large to humans, but do not be fooled. It is but an infinitesimally small pea in the known universe, which has expanded to at least 13 billion light years in all directions since the Big Bang. The extremes of the universe are, thus, thought to be about 26 billion light years apart, which represents a distance of 2.49×10^{23} km or 1.55×10^{23} miles apart. The human imagination is greatly strained to conceive of what such astronomical distances might truly comprehend.

Conclusions

Coping with the dangers of comets and devising protective strategies to mitigate or eliminate them could be perfected in the next decade or two. Perhaps this might start with the U.N.-sanctioned Space Mission Planning Advisory Group (SMPAG), COSPAR (the Committee on Space Research), or other global scientific or collaborative bodies.

We must store and protect essential supplies and information against the worst type of cosmic accident so that humanity could rebuild modern civilization in the worst scenario. This effort has already begun, but more could be done.

The other half of the equation remains to be solved. We must seek more knowledge about the nature of comets and centaurs, their possible angles of attack, their trajectories, and deflection strategies. The problem of orbital space debris may not seem as severe as a strike by a comet or giant asteroid. Nevertheless, this threat becomes more urgent every day. There is a need for technological research and innovation as well as regulatory reform to address orbital debris mitigation. The sooner the better; delays only make the cost of reform and cleanup ever greater.

Finally, more research needs to be done to consider other types of threats to humanity that could come from cosmic hazards. Changes to the Earth's geomagnetosphere and protective shielding, among other processes and concerns, will be addressed in the following chapter.

References

1. Wall, M.: Earth impact: are comets a bigger danger than asteroids? Space.com, 18 June 2014. https://www.space.com/26264-asteroids-comets-earth-impact-risks.html
2. Centaurs. https://astronomy.swin.edu.au/cosmos/C/Centaurs. Accessed 10 Jan 2021
3. Morrison, D.: What would be the environmental effects if the earth collided with a large comet? Scientific American, 21 Oct 1999. https://www.scientificamerican.com/article/what-would-be-the-environ/
4. Lubin, P., Hughes, G.B.: Directed energy for planetary defense. In: Handbook of Cosmic Hazards and Planetary Defense. Springer Press, Cham (2015)
5. Mosher, D.: Probe to another star system. That may be powerful enough to 'ignite an entire city.' Business Insider, 18 Dec 2018
6. Eves, S.: Space Traffic Control AIAA Progress Series. American Institute of Aeronautics and Astronautics, Reston (2017)
7. Madi, M., Sokolova, O. (eds.): Space Debris Peril: Pathways to Opportunities. CRC Press, Boca Raton (2021)
8. Muelhaupt, T.J., Sorge, M.E., Morin, J., Wilson, R.S.: The technical challenges of better space situational awareness and space traffic management. J Space Saf. Eng. **6**(2), 80–87 (June, 2019)
9. McKnight, D., et al.: The fifty most dangerous space debris objects. Sept 2020. https://iafastro.directory/iac/paper/id/55502/abstract-pdf/IAC-20,A6,2,1,x55502.brief.pdf?2020-09-10.09:43:39

8

Solar Storms, Coronal Mass Ejections, and Solar Shields

By funneling charged particles into Earth's magnetic field, CMEs can trigger geomagnetic storms that ignite dazzling auroral displays. But those storms can also induce dangerous electrical currents in long-distance power lines. The currents last only a few minutes, but they can take out electrical grids by destroying high-voltage transformers—particularly at high latitudes, where Earth's magnetic field lines converge as they arc toward the surface.

–William Murtagh

A Carrington-level, extreme geomagnetic storm is almost inevitable in the future. While the probability of an extreme storm occurring is relatively low at any given time, it is almost inevitable that one will occur eventually. Historical auroral records suggest a return period of 50 years for Quebec-level storms and 150 years for very extreme storms, such as the Carrington Event that occurred 154 years ago. Lloyds of London and Atmospheric and Environmental Research Study, "Solar Storm Risk to the North American Electric Grid," 2013

Introduction

If one were to survey the general public about serious concerns for the future, what would they say? Many of the answers would be predictable. Pandemics, climate change, and weapons of mass destruction would very likely head the list. Others with a broader perspective might mention asteroids and comets, genetic engineering, and pollution. Few would raise the issue of solar storms and coronal mass ejections (CMEs).

Solar storms in the form of a CME are of particular concern because of the weakening of the Earth's ozone layer and magnetosphere, which help ward off

J. N. Pelton, *Space Systems and Sustainability*, https://doi.org/10.1007/978-3-030-75735-9_8

a solar storm's ionic blasts. The Earth's natural magnetic protection has weakened due to a magnetic polar reversal, currently confirmed by NASA and ESA to now be underway. The ozone layer has weakened due to the release of noxious greenhouse gases into the atmosphere in yet another climate change-related hazard.

Without these global shields, our world would have little natural protection. Mars has neither of these and lacks a magnetic core dynamo, making it a very dangerous place to live, despite Elon Musk's vision for a human colony on the red planet. A large enough solar storm on Earth could in a few moments wipe out much of the world's electrical infrastructure. This infrastructure, currently valued in the tens of trillions of U.S. dollars, is now vital to modern life, and a world black-out could touch off a global economic recession. Atmospheric scientist William Muragh of the U.S. Government's agency that monitors solar storms has said:

> What's clear is that widespread blackouts could be catastrophic, especially in countries that depend on highly developed electrical grids. We've done a marvelous job creating a great vulnerability to this threat. Information technologies, fuel pipelines, water pumps, ATMs, everything with a plug would potentially be rendered useless. That's going to affect our ability to govern the country [1].

For the first time in the evolution human civilization, it may be possible to design new space systems and technology to protect the Earth from cosmic hazards such as killer space rocks and violent solar storms. Yet such emerging capabilities give rise to fundamental questions. Are the world's political and legal processes ready for such a prodigious undertaking? Would public opinion even support such activities, in terms of the high costs and the needed time and resources? Step one would be to convince the world that the threat is big enough to take seriously.

This chapter addresses the nature of solar storms and the huge collateral damage that can follow from a massive stream of ions zapping thousands of transformers and pipeline controls all over the Earth. There are some remedial actions that can be taken, which are less dramatic that a magnetic space shield, but these steps are also much less effective.

The Dangers of a Coronal Mass Ejection (CME)

The U.S. National Intelligence Council has warned of a great solar storm in its report on possible global "black swan" events. It also warned that this is a future for which humanity is currently ill-prepared to face. If the blast of ions released from the Sun were powerful enough and aimed in just the right direction, it would create enormous damage. Electrical power grids around the world could fail as transformers were vaporized. The controls on pipelines around the world would burn out and fail. Synchronization of the Internet could be lost. Telecommunications, IT networks, and satellites in orbit could fail. In short, much of the world's vital infrastructure that supplies food, water, electricity, gasoline, natural gas, defense alerts, navigational services, and much more would black out. Aircraft with their burned electronic systems could very well fall out of the sky. It would be the start of a dark period for humanity, figuratively, and literally.

As the positively charged ions from the Sun assault the sky, they would light the night up with great brilliance, and the auroral lights would be seen all over the world down to tropical climes. The loss of power and lines of supply, however, would plunge most of the Earth into darkness. A number of people who are directly dependent on electrical power for life support might become the first fatalities, but many more would follow. Over the weeks and months to follow, there would be heavy loss of jobs in regions without light and a rise in crime. The potential losses would extend across computers and data large and small—banking financial records, social security records, electronic deposits for pay checks and credit card payments—and electronics in cars, trucks, and aircraft. The world would be forced to learn just how dependent they were on modern infrastructure.

Urban populations most dependent on these networks would suffer the most. All around the world, machinery and equipment that depend on power would cease to function, as fuel for generators would be in short supply. This in turn would mean that global supply chains would shut down. Over time, many people, especially those who live in cities without direct access to food and crops, would likely begin to starve. A study by Lloyds of London assisted by a team of technical experts estimated that the damage to just the North American electrical grid could cost close to $3 trillion to replace. This study did not seek to estimate the impact that it would have on human life over time without the rapid restoral of power and operational infrastructure [2]. This would be a very bad day for humanity indeed.

And that is not all. A major solar storm that ejects quadrillions of high-speed ions would generate a natural electromagnetic pulse (EMP) when it blasts through the Earth's atmosphere. This creates the exact same effect as if a very powerful nuclear device had been deliberatively exploded in the upper stratosphere. There is no system in operation today that would provide world leaders with instant information on whether or not the event was a solar storm or a nuclear attack. There is danger that a nuclear counterstrike against suspected enemies might inadvertently occur.

For these reasons and more, solar storms and especially CMEs represent a real human threat. They also have a much higher probability of occurring in the modern world than either a major asteroid or comet strike. It has been 162 years since the Carrington event, and the estimates of scientific studies suggest that such events might occur about once every 150 years.

What Is a CME?

The Sun is a massive cauldron of magnetic activity, and when one magnetic loop snaps and breaks apart, that is when a CME occurs. There has been intensive study on the nature of solar storms for many years, both through astronomical observation plus a series of satellites designed for solar examination and measurements. These efforts have provided vital information on the physical and electromagnetic dynamics of the Sun's internal processes. They confirmed that such events ebb and flow in an 11-year solar cycle. The number of these storms that include CMEs increases at *Solar Maximum* and decreases at *Solar Minimum*, then back to Solar Max in the cycle (See Fig. 8.1).

These solar storms include solar flares that release radiation ranging from ultraviolet and X-rays up to the most energetic radiation, gamma rays. About half of the larger flares also release huge amounts of ions in mass ejections from the Sun's corona at enormous speeds, typically exceeding a million miles an hour. The radiation released from solar flares is of concern to human health in terms of skin cancer and genetic mutation. Fortunately, as mentioned, the ozone layer and Earth's atmosphere protect humanity from most of this radiation. The Earth is also blessed to have a protective atmosphere held in place by the world's geomagnetosphere, formed by the Earth's iron core and the magnetized liquid region that swirls around this core. In extreme southern latitudes where the ozone layer is the weakest and the "Southern Anomaly" exists in the Earth's magnetic field, an ozone hole can sometimes form. This allows ultraviolet radiation and X-rays to penetrate down to the Earth's

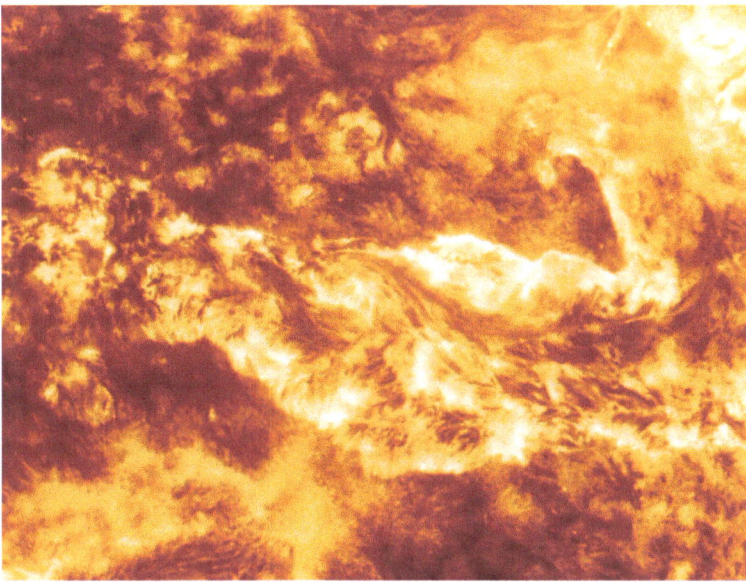

Fig. 8.1 A 200,000-mile-long mass ejection of ions from the Sun's corona, called a CME. (Courtesy of NASA)

surface. In these regions, skin cancer can be elevated and frogs and other amphibians have exhibited genetic mutations.

Records from 1770 in China, Korea, and Japan indicate an event when the skies turned red for 2 weeks in an unusual cosmic event that caused great concern. Today, scientists believe that this might have been a massive CME onslaught [3].

It was in 1859 that a truly massive CME hit the Earth. The start of this event was first detected by a solar astronomer named Richard Carrington—a name we have heard time and again already. On a Thursday morning on September 20, 1859, Carrington noted a strange occurrence on the surface of the Sun. He made a sketch of a huge area on the Sun, covered in sunspots much larger than the Earth. This mass of sunspots when linked together seemed to be shaped somewhat like a gigantic dragon, as viewed from his special solar observation equipment. He called the household to witness this odd phenomenon.

On Friday, the following day, the ions from the Sun, traveling at millions of miles an hour, hit the Earth. The Northern Lights appeared as far South as Cuba and Hawaii. The electronics and paper in telegraph offices caught on fire. In those days, they were the only electronic sites around to be affected. This superstorm has been estimated to have had the equivalent power of 10 billion atomic bombs. Fortunately, the power was packed into atomic ionic particles and not an asteroid or comet, because this would have been a true mass extinction event [4].

In 1989, there was the Montreal CME event. Suddenly, there were eight million people in and around Montreal, Canada, all the way down to Chicago and Philadelphia, without light. Recovery from this event and replacement of transformers the size of a small building cost a great deal of money and time. Figure 2.1 showed the before and after pictures of a zapped transformer in Chicago that occurred on this occasion [5].

A similar CME event known as the Halloween event occurred in late October 2003. This CME effected satellites, GPS services, and the electrical grid in Scandinavia, where significant power outages occurred. It started on October 26, when three large sunspots began to appear in the Sun's corona. The largest of these grew to 11 times the size of the Earth. On October 28, the first eruption occurred. The events were described as follows:

> A very fast moving burst of gas [i.e. ions] and magnetic energy from the Sun's outer atmosphere, known as a coronal mass ejection, in the form of a geomagnetic storm quickly followed, with the storm growing to become the sixth most intense in over 70 years. Less than 24 hours later, the Sun produced another powerful Earth-directed coronal mass ejection with another extreme geomagnetic storm following quickly on its heels. [6]

Much more recently, in July 2012, a solar storm was detected. This blast blew by a location that the Earth had swept through just 9 days earlier. If the Earth had been at that location at the time, the devastation would likely have been massive [7]. And if the 150-year cycle is indeed correct, this type of event is overdue. Of course these days, much more than telegraph offices are at risk.

Monitoring Solar Storms

Of course, defending against solar storms requires detecting them in advance, so that satellites and critical infrastructure can power down before a major magnetic shock. There are currently 19 projects around the world that monitor various phenomena related to what is called "space weather." These solar monitor systems are operated by various defense agencies and civil governmental systems. We will briefly discuss a few of these in this section.

International Space Weather Initiative (ISWI)

The International Heliosphere Year (IHY) that ended in 2007 gave rise to the so-called International Space Weather Initiative (ISWI). The secretariat for this activity is currently operated though an international partnership involving the U.N. Office of Outer Space Affairs, the University of Kyushu in Japan, NASA, and the Bulgaria Academy of Science.

ISWI activities are mainly related to solar activities that impact the Earth's ionosphere. They are listed at the ISWI website: http://www.iswi-secretariat. org/. There is also a periodic newsletter and ISWI workshops, although the 2020 workshop scheduled to be held in India was cancelled due to the COVID-19 pandemic.

The NOAA and U.S. National Space Weather Center

The U.S. National Oceanic and Atmospheric Agency (NOAA) has solar sensing equipment that monitors solar activities, and also operates a dashboard display that provides near-real-time reports on the nature of current solar activity.

This display, which originates in the NOAA facility in Boulder, Colorado, provides updates on solar wind speeds, solar wind magnetic fields, solar X-ray flux, solar proton flux, and geomagnetic activity levels. It provides a color-coded indication of activity levels that goes from green (safest conditions) up to yellow, orange, and then varying levels of red danger alerts. In addition, this useful report provides a comprehensive 3-day forecast of solar weather and geomagnetic conditions [8]. Figure 8.2 from the US National Space Weather Center shows a real-time image of solar radiation as of December 4, 2020. The blue dot in the middle is imposed over the Sun to reveal more clearly the radiation and ion dispersal into space.

As of December 4, 2020, when this part of this chapter was written, space weather was at one of its calmest periods. In the case of CMEs, the result can be dramatically different. Figure 8.3 depicts what a massive CME would look like in terms of its impact on Earth and its magnetosphere.

A massive CME and the natural EMP it creates can hugely reshape the Earth's magnetosphere. The ions compress the field on the one side and extend it on the other. It could extend the magnetic contours away from the planet over 30 Earth diameters out into deep space.

Fig. 8.2 Real-time image of solar radiation from the Sun as of December 4, 2020. (Courtesy of the U.S. NOAA National Space Weather Center)

Fig. 8.3 A depiction of a solar CME hitting the Earth's magnetosphere. The Earth is the tiny blue dot. (Distance between the Sun and Earth not shown to scale. Courtesy of NASA)

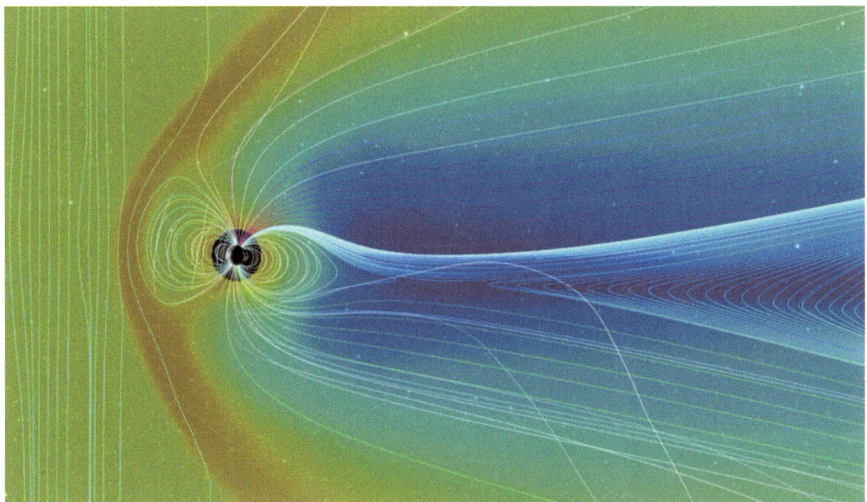

Fig. 8.4   The impact of a CME on the Earth's protective magnetic field. This magnetic field, as depicted in the light blue lines, shields the world and helps retain the Earth's atmosphere. Mars, which has no such magnetic field, is not so lucky. (Courtesy of NASA)

Figure 8.4 is a more scientific representation of what happens. In this instance, the Earth's geomagnetic field facing the Sun is being smashed inward by a gigantic magnetic field of quadrillions of ions, and it thus collapses inward. Meanwhile, the magnetosphere on the other side of the Earth is being violently thrust outward into deep space. The extremes of the magnetic field contours are stretched so far away that they are no longer visible in this image [9].

As noted earlier, Mars, unlike Earth, does not have inner magnetic dynamo occurring within its core to create this global magnetic shielding effect. Without it, Mars also cannot retain an atmosphere, as the solar wind strips the atmosphere away.

Magnetic Polarity and Satellite Constellations

There is increasing concern among some scientists that the Earth might need additional protection due to the threat posed by the magnetic polarity shift that is now occurring. The fear is that a possible giant CME event might occur while this shift is happening, exacerbating the damage wreaked on modern electronics and technologies. The start of such a shift has been confirmed by

Fig. 8.5 The four MMS satellites stacked on top of each other prior to launch as a constellation. These satellites are designed to measure shifts in the Earth's magnetic field. (Courtesy of NASA)

the three satellite ESA Swarm satellite constellation and the four satellites in NASA's Magnetosphere Multi-Scale (MMS) satellite mission (See Fig. 8.5).

These two scientific satellite projects were designed to measure the geomagnetosphere in detail. ESA's Swarm constellation first confirmed the change, and then, NASA' S MMS mission reconfirmed it. Some say the process will take 1000–10,000 years, yet others argue that it might take a 100 years or even less.

Some modeling studies based on data collected by the two constellations have projected that Earth's protective magnetic shielding might be reduced down to only about 10% or 15% of the level being provided today. Figure 8.6 shows on the first panel the normal conditions when the world's

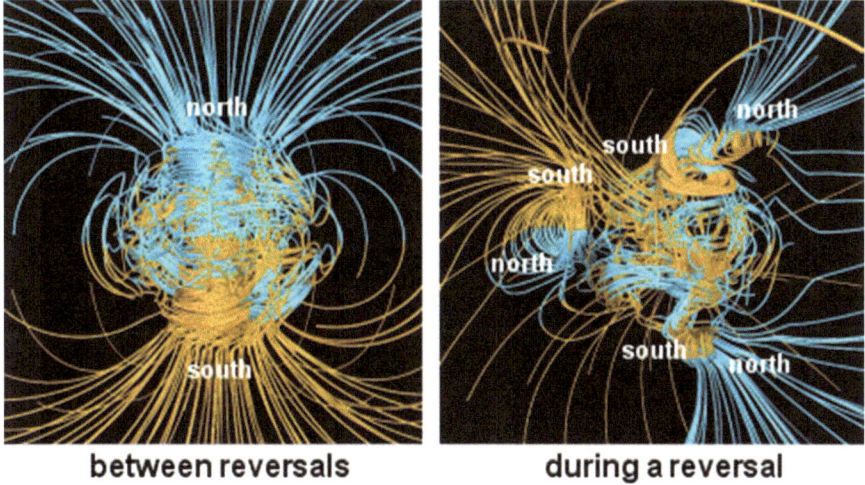

between reversals   **during a reversal**

Fig. 8.6 The Earth's natural magnetosphere protects against violent storms, but this could change during a polarity shift

magnetosphere is in place, while the second panel represents the changes that might occur at the height of the polarity transition. This would not only disrupt the Earth's ability to shield the planet from a major solar storm, but it would lead to other disruptions as well. For one, compasses would no longer work like they used to. In short, this is uncharted territory.

Over time, it is thought that these polarity shifts have occurred over 180 times and averaged about 300,000 years apart, though they happen sporadically and it has been more than twice that time since the last major one, which took place about 800,000 years ago. Exactly why the shift might be occurring now is not known.

In addition to these efforts, it would be useful to widen the purpose and functionality of the International Asteroid Warning Network (IAWN) and the space mission planning advisory group (SMPAG) to include global capabilities and systematic warnings about CMEs.

New Technology to Defend Against Solar Storms

There is a need to retrofit as much electrically powered infrastructure as possible. This would include installing quick-trigger, heavy-duty circuit breakers for all transformers in electrical grids around the world. It would also involve placing as much vital infrastructure as possible on independent, off-grid energy sources not easily interrupted by CMEs. These critical facilities might include alternative backup cooling systems for nuclear power plants, more protected control and communications centers, as well as Internet-related points of presence (POPs—i.e., routers, servers, and other devices that maintain Internet traffic), etc. Where possible, it might be useful to have batteries or other power storage systems that can be recharged and are otherwise disconnected from conventional power sources.

One more exotic idea, known as a LAPSE (LAgrangian Protector against Solar Ejections) would seek to deploy a magnetic shield at Lagrange Point 1, about a million miles out in space. This shield, which would be magnetic rather than physical in nature, would be positioned at Lagrangian Point 1. This is a special location where the gravity of the Sun and that of the Earth essentially cancel each other out. This would therefore be a stable location for the magnetic shield's deployment. If designed properly and operated as planned, it might prove vital to protecting trillions of dollars of vital infrastructure on the Earth's surface and also in Earth orbit. This is a project that could be accomplished for an investment equivalent to many of NASA's prior deep-space satellite missions. It could be one of the most cost-effective missions ever built and deployed. In order to be successful, it would need to a globally engineered and operated system, likely under management of the world's space agencies. This might use mechanisms already established, such as the Inter-Agency Space Debris Coordination Committee, to undertake the planning of such an ambitious project. The actual design and building might perhaps be undertaken as a public–private joint initiative [10].

NASA Chief Scientist James Green has considered this situation in depth and suggested that what is good for the goose might be even better for the gander. His starting premise is that Mars does not have a viable atmosphere for human settlement because it lacks magnetic core dynamo and subsequent magnetic shield, as has been mentioned. Thus, the solar wind strips away the atmosphere as it forms. If a magnetic shield could work to protect the Earth from the solar wind and dangerous CMEs, then it might be possible to create a sort of magnetic field for Mars as well. If this could be sustained for a number of years, this would allow a natural atmosphere to form. Such an approach to terraforming Mars is a rather visionary idea, but it really might be possible from a macro-space engineering perspective [11] (Fig. 8.7).

Another possibility would be to create a Sun radiation screening system at Lagrangian Point 1. This would be a more ambitious project to modulate photons, UV light, and X-rays to ease climate change and global warming.

The idea of creating megastructure shields in space may in 2020 seem like far-fetched science fiction. Yet in the nineteenth century, sending rockets to

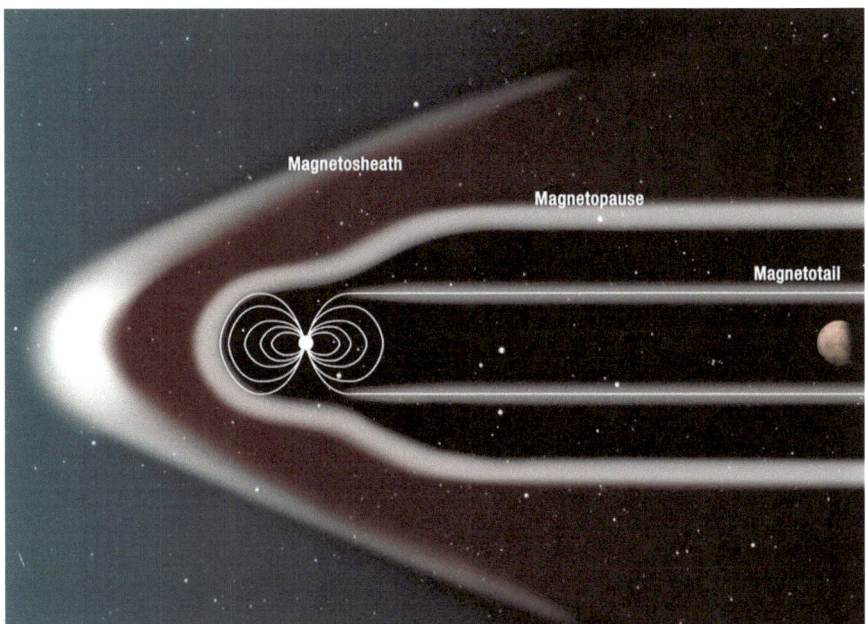

Fig. 8.7 A proposal for a magnetic protective shield for Mars. (Courtesy of James Green of NASA and colleagues)

the Moon and Mars, supercomputers, and the Internet would have seemed preposterous as well. There have been other even more exotic suggestions. One of these, by scientist and science fiction writer David Brin, is to create a steady source of propulsion to gradually move the Earth further from the Sun, protecting the planet from global warming, solar storms, and the expansion of the Sun's radiation over time. Nevertheless, much less exotic space systems and sensors can also be conceived. There could be additional spacecraft systems to simply study the Sun in greater detail.

Conclusions

Solar storms are perhaps the least known and certainly least understood threat among the general public. Yet they are today the most likely cosmic hazard that could create mass global havoc and destruction of vital infrastructure. The factors that make these solar storms of so much concern include the following:

- A global population headed toward 12 billion that is increasingly urban and highly dependent on modern infrastructure operating on electrical power.
- This infrastructure includes satellites, the Internet, pipelines, telecom, computer, and IT networks, defense systems, navigational systems, and electronic automated controls (such as IoT and SCADA systems) for billions of things from elevators to aircraft. The risk extends throughout the world's supply chains, transportation systems, and nuclear power plants. The larger the city, the greater the vulnerability.
- The blast of proton ions from the Sun when it hits the Earth's magnetosphere creates a massive electromagnetic pulse (EMP) that can destroy electrical transformers.
- The current best estimates of the frequency of a dangerous, Carrington-level solar storm hitting the Earth is about every 150 years.
- The current best estimates of the frequency of a Montreal-level solar system hitting the Earth are about every 15–20 years.
- Data from ESA's Swarm and NASA'S MMS constellations show that the Earth's magnetosphere is in flux, and its protective power against a major solar storm is diminishing. At the worst of this transition, it may be reduced to 15% of its current level.
- Strategies for retrofitting the world's electrically powered infrastructure against a major storm have not been effectively implemented.
- Longer term strategies to create solar shields or generate an Earth-based magnetosphere have just been identified, and no serious engineering plans initiated.

The time to initiate serious study is now, in order to save us potential trillions down the line.

References

1. Rosen, J.: Here's how the world could end—and what we can do about it. Science Magazine, July, 2016. https://www.sciencemag.org/news/2016/07/here-s-how-world-could-end-and-what-we-can-do-about-it
2. Solar Storm Risk to the North American Electric Grid. May 2013 Lloyd's of London study http://highfrontier.org/wp-content/uploads/2013/05/Lloyds-Solar-Storm-Risk-to-the-North-American-Electric-Grid-May-2013.pdf
3. Beall, A.: In 1770, a huge solar storm turned the skies of Asia red for two weeks. Wired Magazine, Nov 2017

4. Klein, C.: A Perfect Solar Superstorm: The 1859 Carrington Event. History.com, https://www.history.com/news/a-perfect-solar-superstorm-the-1859-carrington-even. Last accessed 3 Dec 2020
5. 1989 Geomagnetic Storm. https://en.wikipedia.org/wiki/March_1989_geomagnetic_storm. Last accessed 6 Dec 2020
6. Remembering the Great Halloween Solar Storm. U.S. National Centers for Environmental Information. https://www.ncei.noaa.gov/news/great-halloween-solar-storm-2003. Last accessed 8 Dec 2020
7. Solar Storm of 2012. https://en.wikipedia.org/wiki/Solar_storm_of_2012. Last accessed 3 Dec 2020
8. "Space Weather Conditions" Space Weather Prediction Center, NOAA. https://www.swpc.noaa.gov/communities/space-weather-enthusiasts. Last accessed 4 Dec 2020
9. Fox, K.: Modeling Earth's magnetism. https://svs.gsfc.nasa.gov/11689. Last accessed 6 Dec 2020
10. Pelton, J.N.: Saving humanity: Is space up to the job? Room Space J., 20–22 (2018)
11. Green, J., et al.: A Future Mars Environment for Science and Exploration. Planetary Science Vision 2050 Workshop 2017 (LPI Contribution No. 1989)

9

Overpopulation

The constant effort towards population, which is found even in the most vicious societies, increases the number of people before the means of subsistence are increased.

–The Rev. Thomas Malthus

Too many cars, too many factories, too much detergent, too much pesticides, multiplying contrails, inadequate sewage treatment plants, too little water, too much carbon dioxide—all can be traced easily to too many people.

–Paul Ehrlich, Author of The Population Bomb

Introduction

Of all the pundits, visionaries, and seers of the past century, Arthur C. Clarke stands out above them all. The so-called "Oracle of Colombo" was a prodigious source of predictions about not only communication satellites and remote sensing satellites but also everything from the Internet to vaccines, DNA testing, artificial insemination and birth control, artificial intelligence, driverless cars, and much more.

Clarke's Tele-Cities

Clarke foresaw a future where virtual reality-based telecommunication systems, energy efficiency, and tele-services could lead to a world of tele-cities. This would be a future in which most jobs and professions involved tele-services. In this world, the nature of urban architecture could be different, and much less people would be needed to operate a functioning society. Most farming and manufacturing would, for instance, be the purview of smart robots. In such a world, healthcare, education, and training, going to work, and shopping could in many instances be accomplished electronically. It is easier and more energy efficient to move electrons than people [1]. In Clarke's

© The Author(s), under exclusive license to Springer Nature Switzerland AG 2021
J. N. Pelton, *Space Systems and Sustainability*, https://doi.org/10.1007/978-3-030-75735-9_9

view, future cities could be more energy efficient, housing less expensive, and citizenry better distributed, less dense, or overcrowded.

The logic of this idea has taken some time to evolve. And clearly, a personal touch must still be a part of a meaningful human future. The prime force that today is promoting Clarke's idea of the tele-city comes from fear of infection and the COVID-19 pandemic. This is a time where telecommuting for work, tele-health, and tele-education can save lives from the ravages of pandemics. The fringe benefits of such a future, however, could include the fact that it also saves energy, reduces air pollution, and saves personal time that would otherwise be spent commuting.

New Ways of Thinking About the Future

Paul Ehrlich's so-called "population bomb" is still rapidly exploding, especially in lesser developed economies. This human overcrowding is increasingly straining the Earth's resources and giving rise to faster spreading forms of zoonotic diseases.

The hope is that new space technology for remote sensing and very sophisticated data analytics can be used to measure the growth of human population and its impacts on the planet with much greater accuracy. As previously discussed, it is also hoped that related analytic and enhanced remote sensing processes can identify areas where the latest animal-to-human viruses have occurred and contain their spread. Perhaps other tools can be used by global leaders to create incentives for limiting population growth around the world.

This chapter examines how human population growth represents a major risk to long-term survival of life on planet Earth. Limiting this growth is in many ways the key to the future. Effective use of tele-city concepts, smart information, data analytics, and space technologies might become part of a new way of life based on a future sparser human population.

Overpopulation and Pandemics

A recent article in the June 2020 edition of *Population Connections* was entitled: "Destruction of Habitat and Loss of Biodiversity Are Creating the Perfect Conditions for Diseases like Covid-19 to Emerge." This analysis suggested that pandemics are becoming more frequent in part because of human disruption of the world's natural biosystems. The problem has been described this way: "We cut the trees; we kill the animals or cage them and send them to markets. We disrupt ecosystems, and we shake viruses loose from their natural hosts. When that happens, they need a new host. Often we are it" [2, p. 15].

The suggestion is that the start of the COVID-19 virus should not simply be attributed to a bat or a pangolin in China. More overarchingly, it is the result of human population growth that disturbs natural ecosystems and increases opportunities for viruses to jump from animals to humans. The National Center for Emerging and Zoonotic Diseases and the Centers for Disease Control has now stated that "Each year around the world, it is estimated that zoonotic diseases … cause 2.5 billion cases of sickness and 2.7 million deaths" [3]. This same analysis suggested that with current population growth and global transportation systems, the problem will get progressively worse.

The human population forecast for the world by 2100 is variously estimated to be between 10 and 12 billion. Such a population increase will portend a very dangerous future. Dr. Peter Daszak is President of the EcoHealth Alliance, a non-profit specialist in tackling wildlife-borne disease. He reported in a study that more than 300 new infection diseases were identified between 1940 and 2004. Dr. Daszak explains: "These were either Zoonoses (those that jumped from animals to humans) or a new pathogen variant that had become resistant to available drugs. This occurred during the period when world population increased by over 3 billion people and significant new habitats were invaded" [2, p.12].

The U.S. Agency for International Development program, known as "PREDICT," began in 2009 under the administration of George W. Bush and continued under the administration of Barack Obama. Aimed at detecting zoonotic diseases, the program uncovered some 160 pathogens that had to potential to evolve into pandemic viruses. It was shut down in September 2019 by the Trump Administration. Dozens of experts operating in Asia, including China, were disbanded just months before the COVID-19 pandemic erupted. "It was a genius, visionary program that USAid took a big risk to fund and it's a crying shame it was canceled," said Dr. Daszak, whose organization was one of the major partners in the program [4].

Now is the time for a more systematic approach to sustaining life on Earth. This would include:

 (i) Reactivating programs like the USAID "PREDICT"
 (ii) Creating global funds to limit incursions into jungles and wildlife habitats
(iii) Developing lower cost birth control technology
(iv) Providing global incentives to reduce population growth rates

Other Overpopulation-Related Issues

The famous eighteenth-century economist Thomas Malthus well understood the problem of population growth. The mathematics he considered in his 1798 analysis remain even truer and more powerful today: "Population, when unchecked, increases in a geometrical ratio. Subsistence, increases only in an arithmetical ratio" [5]. Other processes function in a similar way. The effects of pollution can compound. Oil spills have an adverse effect on sea and shore life, impacting the production of oxygen and absorption of carbon dioxide. The residue drifts to the icecaps, freezes into the icy areas, and then changes the albedo. This quickens the thawing process. The water that thaws becomes salt water. Refreezing does not easily occur. It is many more times easier to pollute than to reverse the pollution process.

Potable water is becoming the most powerful and scarce resource for human sustainability. This reality was apparent to this author in 2018 while teaching satellite systems to students at the University of Capetown. At that time, the city's water supply for a city of over two million was running dry, and "Day Zero" loomed. This referred to the time when the water levels might fall too low, ushering in Level 7 restrictions in which the municipal water supply was turned off, and the citizens of Capetown would have to queue for daily water rations. Eventually, the city recovered before reaching this crisis point, but nevertheless, the author's short 2-week stay in the region made the issue of too many people and too little water all too real [6].

By 2100, the world will likely be 80%–85% urban. We will likely see the incredible growth of megacities, defined as urban areas of over ten million people, as well as very large cities with populations in the range of 5–10 million inhabitants. The latest U.S. National Intelligence Council's Global Trends projects that the number of megacities (i.e., over ten million) will grow from 28 as of 2014 to 45 megacities by 2030. The number of cities with populations in the range of 5–10 million will in the same time period increase by 20 cities, from 43 to 63. Medium-sized cities from one million to three million are expected to increase in number by 141 cities, from 417 to 558. Most of these will be in Africa and South Asia. This same report estimated that as of 2030, there will be 5.1 billion people living in cities out of a total world population of perhaps 8.5 billion or more. This would be 60% urban, up from about 53% urban as of the end of 2020 [7].

Put bread, sugar, and water inside a closed test tube, and bacterial organisms exponentially multiply—until they do not. We need to find a way to live within the resource limits of Spaceship Earth. It beats the alternative.

The Case Study of Nigeria

Arthur C. Clarke was noted for the following witticism: "Forecasts are always difficult, especially about the future." This joke was casually mentioned during a chat about future projections in the upstairs library room of Clarke's villa in Colombo, Sri Lanka. The implication was that public officials often looked at outcomes and then said, this is just what I predicted all along!

Rather than relying on forecasts of the future, it is more useful to examine the hard facts. Data and case studies are certainly a useful way to understand powerful demographic trends. Therefore, let's examine the experience of Nigeria.

Table 9.1 provides the data from the world population review database. These figures show that the countries around the world with the highest birth rates are all from Africa. This is in part because most African countries are still heavily agrarian, and thus, larger families are seen as necessary to support

Table 9.1 The top 25 countries around the world in terms of birth rate

The 25 countries with the highest birth rates		
Rank	Country's birth rate	Annual percentage
1	Niger	7.153
2	Somalia	6.123
3	Dr Congo	5.963
4	Mali	5.922
5	Chad	5.797
6	Angola	5.589
7	Burundi	5.577
8	Uganda	5.456
9	Nigeria	5.417
10	Timor Leste	5.337
11	Gambia	5.318
12	Burkina Faso	5.231
13	Mozambique	5.143
14	Tanzania	4.924
15	Zambia	4.901
16	Benin	4.867
17	Ivory Coast	4.811
18	Central African Republic	4.754
19	Guinea	4.738
20	South Sudan	4.736
21	Senegal	4.647
22	Cameroon	4.603
23	Mauritania	4.576
24	Republic Of The Congo	4.561
25	Equatorial Guinea	4.554

Source: World Population Review

farming labor requirements. It should also be noted that these countries also have the highest death rates, so the net total population growth for most of these countries is around 3–4% [8].

Nigeria is actually not the country with the highest birth rate in the world—that title belongs to Niger. Yet Nigeria is the most significant because it is the largest country in Africa and its population growth in absolute numbers is by far the largest. For the last decade, Nigeria has had an annual net growth rate of around 3.0% and an annual birth rate of around 6%–7%. This net growth of 3% may not seem like a lot, but it is a very worrisome number indeed.

An article on Nigeria's population growth has summarized the situation as follows:

> A UN report last year projected that, by 2050, Nigeria will become the world's third largest country by population and one of the six nations with a population of over 300 million. The country, which currently has a population of just under 200 million, will contribute a significant amount to the 1.3 billion people projected to be added to Africa's current population in the same time frame [9].

Nigeria, along with its seven cities with populations over a million, has had a very difficult time building enough services and infrastructure to keep up with such a population surge. Between 1960 and 2021, Lagos, Nigeria, grew from 763,000 to 13 million. This is probably the largest expansion of a major city anywhere in the world. The explosive rate of expansion would be overwhelming for any city, and the staggering growth was certainly a factor in moving the national capital from Lagos to Abuja as of 1991.

A 2018 report of the World Bank on Nigeria's economic development notes a growth in urbanization from 41% to 50% in the decade between 2007 and 2017. A more recent report described conditions as follows: "The realities of life in Nigerian cities are hard. In Lagos, about two of every three people live in a slum. Less than 10 percent of residents have access to piped water (for those that do, it is often riddled with sediment and unsafe to drink)" [10].

On New Year's Day of 2018, Nigeria recorded 20,210 new babies born in a country of nearly 200 million. In contrast, China—a country with a population that is seven times larger—had only 44,760 births that day. Thus, on a per capita basis, China had 3.5 times less babies born than in Nigeria. That day, India recorded the birth of 69,040 babies. On a per capita basis, this was 2.6 times less babies born than in Nigeria on a per capita basis. In the last 2 years, U.N. projections indicate that the birth rate in Nigeria has declined slightly, but not enough to stem a giant increase in population.

Perhaps the largest issue of Nigeria's population boom is that today, the population is composed of a huge number of young people that put a strain on its economy. Nearly 44% of the population are 14 or under, and 63% are under 25. Only about 7% are 55 or older. This creates enormous problems and stress on schooling, healthcare, housing, etc., along with an undersupply of workers, taxpayers, and adults [10].

China and Singapore

China has blossomed economically over the last 40 years. During this time, the per capita income increased by a remarkable factor of 16. Key aspects of this growth were of course industrial growth and global trade, but also the shrinking size of families. Policies that imposed delayed dates for marriage and a one-child policy were harsh, but they did in fact control rapid population growth.

Many people also point to the remarkable economic miracle of Singapore. There was incredible expansion under the leadership of Prime Minister Lee Kwan Yew. Again, part of that trend was a tax policy that discouraged population increase. Under Yew's tax policy, a family got a tax credit for one child, but lost its tax credit for having two children and was penalized for having more than two children. Under this policy, Singapore achieved a zero population growth and its wealth increased.

As India overtakes China as the world's largest country, so too will Nigeria overtake the United States as the third largest.

Limits to Growth and Species Extinction

There are those that say continued population growth achieves ever bigger and better economies. Economies have continued to grow. Industrial production has managed to expand. There is always a way to feed more people and meet their needs for water and resources. So why should stop now?

As human populations claim more land and disrupt more natural habitats, animal extinctions increase and vegetative diversity shrinks. The International Union for Conservation of Nature (IUCN) has evaluated some 5850 species classified as mammals. After detailed evaluation, it found that 1244 on the low end and 2116 on the high end of those mammals were considered critically endangered, endangered, or vulnerable. In addition, 81–83 were considered extinct in the wild.

Species Extinction and Human Population

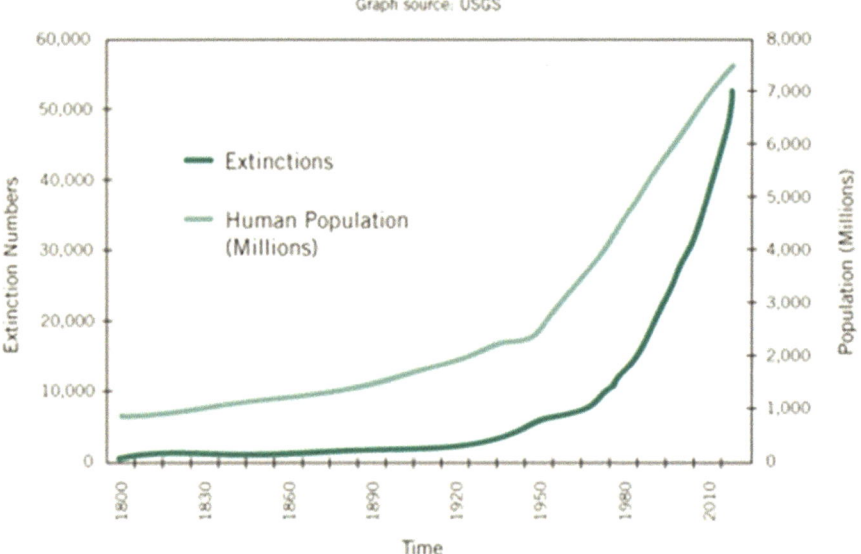

Graph source: USGS

Fig. 9.1 The rise of human populations and mass species extinction in the last centuries (Courtesy of the Convention on Biological Diversity (CBD) at commondreams.org)

Figure 9.1 shows the correlation between the growth of the human population and the number of wildlife extinctions that are now occurring. The last century has made an enormous impact. The twenty-first century will be far more severe. In the twenty-second century, human extinction may become the prime concern.

Humans are, in the words of scientist, inventor, and writer Neil Ruzik, "big eaters." Those in the economically developed world consume the most. These privileged people eat the most calories, use the most plastics, paper products, gasoline, electrical energy, potable water, and have the largest houses to heat and cool.

As ocean temperatures rise, ocean-based coral reefs—key to breeding the fish we eat and sustaining diverse ocean life—are dying off. This change is also causing the die-off of algae and plankton, which as we have seen produce oxygen and remove carbon dioxide from the atmosphere. In South America and Africa, rainforests are being lost to large swaths of domesticated crop fields. Elephants are killed to create ivory jewelry and whales are killed for oil, baleen, ambergris, meat, candles, and soap, along with many other "exotic" animals used for high-end products.

Table 9.2 Assessment of the International Union for Conservation of Nature (IUCN) on the rapid decline of many species

Current or projected threat of extinction or severe losses	Current threat level
Bird Species	1 out of 8
Mammals	1 out of 4
Amphibians	1 out of 3
Marine Turtles	6 out of 7
Conifers	1 out of 4
Reef building corals	1 out of 3
Agricultural crops	75% of genetic diversity has been lost
Scarcity of water supply for people	Currently 350 million people
If global average temperature rises by more than 3.5 ° C/6.3 ° F	70% of all species risk extinction (Within a century)

Source: International Union for Conservation of Nature
https://www.globalissues.org/article/171/loss-of-biodiversity-and-extinctions#MassiveExtinctionsFromHumanActivity

The IUCN prepared the data presented in Table 9.2. The reality is stark: rapid human population growth could lead to the mass extinction of animals within a century.

The dangers are not just for animals of vegetation any more. As we have seen, the scarcity of potable water is becoming an increasing problem. While water covers a good portion of the planet's surface, the actual amount of water by volume in relation to the overall volume of our planet is small (see Fig. 9.2). Of this amount, less than 1% is accessible freshwater, which must meet the needs of animals, vegetation, and humans.

In a time of rising global population, pollution, and consumption of resources, the problems of a disposable rather than recyclable or sustainable economy are compounded. Urgent action is needed to prevent an Anthropocene extinction event. Fortunately, there is a range of new tools that allow humans to understand more clearly their overall impact on the planet day by day and even minute by minute.

New Space Tools and Data Analytics

There are now many remote sensing space systems that can precisely inventory the growth of cities and human settlements. Landsat satellite imaging analytics have been used for some time to provide more accurate total population and population growth figures [11]. New planned systems such as the

Fig. 9.2 The small volume of water on Earth compared to the overall size and volume of the planet. (Courtesy of the Sierra Club)

Theia network, and existing systems such as Planet, can provide day by day profiles of the changes.

Key information collected from these satellites includes the following:

- The growth profiles and sizes in urban areas around the world [12].
- Changes in areas used for farming and agriculture, and especially loss of areas previous reserved as rainforests or for forestry
- The global mapping of pollution across land, sea, and the atmosphere [13].
- Changes in rivers, streams, water reservoirs, lakes, and ocean coastlines
- Detailed mapping of ocean acidity, algae and plankton health, thermal changes, coral reefs, fish and sea life population densities, red tide blossoming, and other indicators

- Detailed mapping of the arctic regions
- Detailed mapping of the world's flora and fauna and any disruptions to them, including wildfires, crop or forestry disease, volcanic eruptions, etc.

These are only some of the space-based analytic tools that can be used to measure the impact of human populations on the planet. The above cited examples have already in part been accomplished with remote sensing satellites that are essentially obsolete, in comparison with the new satellites that will be deployed in the 2020s and their high-speed computer analysis capabilities.

There are new satellite systems that are being deployed with twenty-twenty vision. Instead of "seeing the world" with a very large overview—with something like 20–5 m of resolution (i.e., possibly able to detect a football field or a canyon on the ground), these new satellite networks have a much keener eye. They will be able to "see" things with up to 20 cm or about a half foot of spatial resolution. Thus, these satellites could detect the location of a person or large animal, or tell what kind of a vehicle is in the image. When these types of remote sensing satellites are deployed, entirely new capabilities become possible. The key becomes connecting this great mass of data in a comprehensible way.

These new satellites will thus be able to discern different types of vegetation, weeds, or crop blight with precision and sound almost instant alarms. Some systems will be able to instantly form a complete image of the entire Earth with an input of 2 quadrillion (that's 2000 trillions) or 2 petabytes of information per global image every 60 seconds. This will reveal the world instant by instant everywhere there is not cloud cover. In contrast, 5G broadband information systems using IoT technology will be able to cover only about 9% of the world's land area. Clever programming will be able to accurately chart trends and changes around the world, allowing a global vision of every bridge, road, house, building, railway system, farm, and indeed every tree on the planet. In a world where information is power, these systems will be powerful indeed.

The Lessons from Biosphere II

Some say that the Biosphere II experiment carried out in Arizona in the United States some decades back was a total failure. To some, it was a stunning success. It proved, at least then, that humans were incapable of creating a sustainable world.

Taber McCallum, a student at the International Space University who later served on its Board in the mid-1990s, was one such biospherean. Taber married Jane Poynter, who had also been in Biosphere II as they struggled to survive. The participants sought to raise corn and other crops that would create oxygen in order to cope with the carbon-based gases (i.e., CO_2 and methane) produced by the biosphereans and other biological processes occurring within the domed structure. Eventually, it was decided to vent carbon dioxide from the exotic structure into the Arizona Desert. Without a release of greenhouse gases, everyone inside Biosphere II would have died.

Taber explained:

> …The Biosphere II experience also altered my life and led to my marriage, but its most profound impact was in many ways just the reverse. I realized how tenuous life could be on planet Earth, and how little was known about how to truly sustain life for the longer term. Before they decided to vent the carbon dioxide there were times when we were gasping for breath. The biosphereans even divided into tribes. Some of the best and most cherished aspects of civilization broke down as there became a true struggle for survival. As the days, weeks, and months went by it got worse and worse.[1]

It might be instructive to put the political and business leaders of the world into Biosphere II until they agreed on policies to reduce carbon-based pollutants and adopt reasonable population control measures.

Conclusions

The swelling world population is becoming an ever greater problem. Spaceship Earth has finite resources. The laws of physics indicate that continued exponential growth is not consistent with human survival.

What are the exact limits to growth? These are not known. Yet the so-called Anthropocene extinction event continues to be driven forward by population increase perhaps more so than any other factor. Conservation and clean energy alone are not sufficient to put the brakes on the ill effects of continued population growth. Our world is increasingly based on a service economy rather than an agricultural or mining economy. By the middle of 2050, most of the world's farming, mining, truck driving, industrial manufacturing,

---

[1] Quote from Taber McCallen, former biospherean, who spent many weeks inside Biosphere II, Arizona, USA.

merchandising, and many other jobs will be done by smart machines.[2] Every logical examination of the global economy and a declining need for humans in the world workforce suggests that population growth must be curbed.

Many individuals have at different times across the centuries explained the need for limits to growth. Yet their warnings have been largely ignored. Reform is so very difficult among a global population of humans spread among some 200 countries and territories. It is compounded when leaders of major nations deny the nature and degree of the problem. The danger is seen only in the form of climate change, but it is in fact a complex stew of inter-related problems.

Despite much empirical evidence and scientific fact that prove Spaceship Earth's physical limits, it is hard to get consensus on almost any particular course of action concerning the core issues at stake. Disputes continue about the wisdom or need for urgent action to conserve nature, control population growth, and reduce atmospheric, ocean, and land pollution. Opposition abounds—especially if such reforms would hurt particular jobs or industry, slow economic growth, or reduce the number of consumers.

References

1. Pelton, J.N.: The Oracle of Colombo: How Arthur C. Clarke Revealed the Future. Arthur C. Clarke Foundation, Washington, D.C. (2015)
2. Vidal, J.: Destruction of habitat—loss of biodiversity. Population Connection, June 2020
3. Pandemics: one health. Population Control, June 2020, p. 5
4. Milman, O.: Trump administration cut pandemic early warning program in September. The Guardian, 3 Apr 2020. https://www.theguardian.com/world/2020/apr/03/trump-scrapped-pandemic-early-warning-program-system-before-coronavirus
5. Malthus, T.: Britannica. https://www.britannica.com/biography/Thomas-Malthus. Accessed 18 Sept 2020
6. Becker, R.: Today wasn't day zero in Cape Town but the water crisis isn't over" The Verge, 11 May 2018. https://www.theverge.com/2018/5/11/17346276/day-zero-cape-town-south-africa-water-shortage-reservoirs-dams-climate-change
7. Global trends: paradox of progress, U.S. National Intelligence Council. https://www.dni.gov/files/documents/nic/GT-Full-Report.pdf. Accessed 15 Dec 2020
8. Total fertility rate 2020. https://worldpopulationreview.com/country-rankings/total-fertility-rate. Accessed 15 Sept 2020

---

[2] More than 60% of Nigeria's population is under 25, https://theatlas.com/charts/rk2o5ocF.

9. Kazeem, Y.: Nigeria's population problem is the result of poor policy implementation. Quartz Africa, 5 Jan 2018. https://qz.com/africa/1171606/nigeria-population-growth-rising-unemployment-and-migration-suggest-things-could-get-worse/
10. Campbell, J.: Home to half the population, Nigeria's cities continue to boom, 22 Aug 2019. https://www.cfr.org/blog/home-over-half-population-nigerias-cities-continue-boom#:~:text=In%20Nigeria%2C%20this%20can%20be,the%20population%20reaches%2021%20million
11. Hillson, R., et al.: Estimating the size of urban populations using Landsat images: a case study of Bo, Sierra Leone, West Africa. Int. J. Health Geograph. 11 July 2019. https://ij-healthgeographics.biomedcentral.com/articles/10.1186/s12942-019-0180-1
12. Al-Bilbisi, H., et al.: Spatial monitoring of urban expansion using satellite remote sensing. https://www.researchgate.net/publication/332435773_Spatial_Monitoring_of_Urban_Expansion_Using_Satellite_Remote_Sensing_Images_A_Case_Study_of_Amman_City_Jordan. April 2019
13. Themistocleous, K., et al.: Applications of satellite remote sensing & GIS to urban air-quality monitoring: potential solutions and suggestions for the Cyprus area. https://www.researchgate.net/publication/257067287_Applications_of_Satellite_Remote_Sensing_GIS_to_Urban_Air-Quality_Monitoring_potential_solutions_and_suggestions_for_the_Cyprus_area. Accessed 29 Oct 2020

10

Climate Change

There are no passengers on Space Ship Earth. We are all crew.
    –Marshall McLuhan

Population growth is 'fueling' the need for more and more cars, houses, and buildings that still consume more carbon-based fuels. Many companies still do not have a clue as to how much greenhouse gas they emit, and most have not yet acted to create and implement viable plans to curb these emissions. Advances in photovoltaic solar cells, electric cars, and high performance batteries fall far short of the Paris Accord objectives for a 'new technology framework.'
    –Joseph N. Pelton, Preparing for the Next Cyber Revolution

Introduction

On November, 4, 2016, the United Nations Paris Accords, known as the Framework Convention on Climate Change (UNFCCC), went into effect. These landmark accords had the support of virtually all the nations of the world and most importantly were backed by the countries that contribute the most to global pollution. Many of the other signatories represented the countries with the most to lose by runaway climate change.

The accords were initially signed by 196 countries. Currently, some 200 countries have now ratified and are backing this convention. The only two countries with significant amounts of pollution on the global stage that are not signatories to this agreement are Turkey and Iran. At least, that statement was true until November 4, 2020. This was the day when the U.S. President Donald Trump made good on his decision to have the United States formally withdraw from the Paris Accords.

The hundreds of revised U.S. environmental policies and the gutting of clean air and water standards adopted during the Trump administration have been of enormous concern around the world. These changes have included relaxing environmental regulations, cutting or limiting procedures to reduce greenhouse gas pollution, and not making contributions to funds that help developing countries fight climate change. The heads of the U.S. Environmental Protection Administration (EPA) have frequently advocated policies that seemed closely aligned with the oil and carbon-fuel industries.

J. N. Pelton, *Space Systems and Sustainability*, https://doi.org/10.1007/978-3-030-75735-9_10

The chair of the Paris Accord talks, Ambassador Laurence Tubiana of France, stated just prior to U.S. withdrawal, "It is clear President Trump has slowed down global progress and ambition on climate since taking office in 2017, and if climate deniers keep the White House and Congress, delivering a climate-safe planet will be slower and more challenging" [1]. Fortunately, Ambassador Tubiana's worst fears were not realized, and the U.S. environmental policies are reversing course once again under the Biden administration.

These events are quite important from a global perspective. The United States is currently responsible for about 15% of the production of greenhouse gases. It is second only to China in terms of greenhouse gas emissions and other activities that pollute the ocean and increase its acidity. Even with the reduction of automotive pollution due to the pandemic, the United States now seems unlikely to meet the Paris Accord goals to lower carbon emissions to a level that is 26–28% below 2005 levels by 2025 [1].

The election of President Biden represents a key turnabout on U.S. policy. Only a few months after the United States withdrew from the Paris Accords, on his first day in office, President Biden signed an executive order to rejoin the agreement. He also tapped former U.S. Secretary John Kerry as special ambassador to pursue U.S. commitments on climate change reforms. Ambassador Kerry actually played a key role in the negotiation of the Paris Accords. Biden's appointment of Michael Regan to head EPA and Deb Haaland to head the Department of Interior are strong indications of a concerted effort to reverse efforts to derail environmental reforms both in the United States and globally. Nevertheless, time and remedial actions have been lost. Further, the echoes of climate change deniers remain to be addressed [2].

The setbacks in meeting the goals of the Paris Accords are not restricted to the United States alone. Many other so-called "populist" leaders that have been elected in Brazil, Turkey, Eastern Europe, and elsewhere in the world have also found it convenient to change course as President Trump did. As discussed in Chap. 4 on pollution, these changes have brought on rising environmental concerns. Some 10% of the Brazilian tropical forests have been lost since the election of far-right candidate Jair Bolsonaro as President in 2018, along with his aggressive campaign to add new farmland in the Amazon.

The fundamental weakness of the Paris Accords is that the nearly 200 countries that are now within Framework Convention are only "working towards" their indicated goals. There are no enforcement mechanisms to ensure that pledges are actually being met. The lack of firm, mandatory commitment by countries to meet their goals for significant greenhouse emission reductions indicates that much more still needs to be achieved. Fortunately, the latest space remote sensing systems could help enforce any mandatory objectives that might be established.

Climate change problems are not a far off concern for some nations, who see action to curtail climate change as vital to their survival. The 44 members of the Alliance of Small Island States, which includes countries such as Tuvalu and the Seychelles, are literally fighting for their lives. They see rising ocean levels as true threats to their nations. And for such groups, the unreliable role of the U.S. Government has become a focus of concern. The lead negotiator for the alliance, Carlos Fuller, has noted with some despair the alternating role that the U.S. Presidents have played over the years with an almost bizarre "yo-yoing" set of positive and then negative policies.

It was in fact the United States that led the efforts to bring international climate change agreements into force (i.e., both the Kyoto Protocol and the Paris Accords). Yet it was subsequently elected U.S. presidents that then tried to stop these agreements from coming into effect. After the departure of the United States from the Paris Accords, Fuller said:

> I am also reminded of Kyoto because I see the ways history threatens to repeat itself. I am not superstitious, but for us at the Alliance of Small Island States, who have been in the trenches of this climate fight for 30 years, our memory is long. The US was instrumental in negotiating the Kyoto Protocol, only to fail to ratify it at the moment of truth; a blow the treaty never really recovered from. It is doing the same with the Paris Agreement [3].

Fuller goes on to explain that the world community, unlike in 2010, is now looking to others to provide leadership and constancy of purpose, and that the Paris Accords do not need the United States to proceed. Indeed, he said: "China is positioning itself to fill this leadership void" [3].

The good news is that billions of dollars in new governmental programs as well as bonds to support Paris Accord environmental goals have been authorized by legislative measures and executive action. Further, unlike the situation with the Kyoto Protocol, no other nation has followed the U.S. lead and withdrawn from the Paris Accords. Even in the United States, a "We Are Still In" campaign was organized by the World Wildlife Foundation, and some 4000 corporate executives, mayors, governors, tribal leaders, and others signed up to honor the Paris Accords. These leaders sought to meet U.S. objectives as originally made. The WWF-led coalition is thus following a bottom-up approach, representing $9.5 trillion of the U.S. gross domestic product, or over 70% of the U.S. economy [4]. Due to these efforts, significant environmental progress in the United States was still being made despite the U.S. government's temporary departure.

The latest conference for nations to update their commitments and review global progress was canceled due to the Covid-19 pandemic but will now be held in 2021.

The United Nations Sustainable Development Goals

Efforts to create a sustainable planet go beyond the Paris Accords. The United Nations and the United Nations Environmental Programme have been concerned with climate change, global pollution, greenhouse gases, ocean acidification, and loss of native habitats around the world for years. There are notable U.N.-led efforts that are seeking a top-down approach to global goals, hoping to make progress in many areas that relate to climate change.

The U.N. Goals for Sustainable Development were introduced and mentioned several times throughout this book (see in particular Chap. 2, and Fig. 2.1). The objectives that are most relevant to this environmental and sustainability effort include:

- Climate Action (Goal 13)
- Life Below Water (Goal 14)
- Life on Land (Goal 15)

There was also specific targeted action that addressed Clean Water and Sanitation (Goal 6), Affordable and Clean Energy (Goal 7), and Sustainable Cities and Communities (Goal 11). All six of these specified goals addressed ways to make Earth more sustainable, less polluted, and more habitable. Other goals for Good Health and Well Being (Goal 3) and Quality Education (Goal 4) indicated how improvements in these areas, like a curriculum that stressed a cleaner and more sustainable world, could lead to improvements in the environment.

Goal 6 for Clean Water and Sanitation This goal set eight specific targets that were easy to comprehend. Yet there were no means identified as to how changes to water supplies and sanitation systems could actually be designed and implemented by countries within 15 years. Likewise, there was no mention of specific budgets or financial mechanisms to pay for these water projects, nor how they would be globally monitored to see whether they had truly been achieved. Goal 7 for Clean and Affordable Energy This goal established five specific targets for more accessible, modern, cleaner, and internationally backed innovation in the supply of energy. Again, the hows and whys were not specified. The problem of reducing fossil fuels while increasing clean energy was clearly stated. But how the coal-fired power plants would be phased out, replaced, and new plants paid for was again not explained. Goal 11 for Sustainable Cities and Communities This goal involved ten target areas to make cities and human settlements inclusive, safe, resilient, and sustainable. The problem is that some of the targeted aims might be seen as working against each other. For instance, long-term sustainability might be at times be sacrificed to making a city economically resilient. Goal 13 for Climate Action This goal set five targets for improved climate action. Yet the United Nations also noted the difficulty of the challenge. The current mean world temperature is projected to increase by 3.2 ° C or 6 ° F by the year 2100. The global economy is currently investing $781 million in fossil fuels, which is $100 billion more than the $681 that is being invested in global climate finance. This is currently not an equation for success. Goal 14 for Life Below Water This goal recognized a specific number of challenges and set five targets. These addressed making progress to reduce greenhouse gases and subsequent acid rain. Targets were also set to address illegal fishing, other forms of ocean pollution (especially plastic poisoning and oil spills), and efforts to reverse the increase in ocean temperatures. Goal 15 for Life on Land This goal included 12 targets and perhaps involves the most complex form of human behavior reversal. It focused on sustainable uses of ecosystems, reducing desertification, halting and reversing land degradation, and reversing the loss of biodiversity [5]. The greatest problem in achieving these U.N. goals is the significant diversity of needs, practices, and degree of industrialization of the over 200 committed countries around the world today. This disparity also leads to a huge disparity in the types of environmental challenges to be faced. In short, one size does not fit all. How environmental resources are husbanded and the urgency that countries attribute to fighting climate change and addressing global pollution may greatly differ from country to country, based on their current state of economic development as well as their ability to enforce cleanup or environmental protection measures.

True climate change progress involves environmental reforms with firm objectives against deadlines backed by the "teeth" of enforcement powers. National laws are backed by police forces, a judicial system that can impose fines or jail sentences, and a legislative process that can create new reforms.

They can even sanction new agencies that mandate environmental standards and regulations, spelling out what citizens, corporations, organizations, and entire societies must do. In the case of shared global resources—sometimes call the *global commons*—we lack such enforcement bodies. For the seas and oceans, the breathable atmosphere, the stratosphere, the geomagnetosphere, radio frequencies, and the world's tropical rainforests, we might negotiate treaties and conventions, but these are not the same when it comes to regulatory oversight.

As colleague Chris Johnson, an international space lawyer with the Secure World Foundation, has observed:

> We find ourselves somewhat at a crossroad. There is much at stake here. We must avoid a "tragedy of the commons" type scenario. We need to develop a regulatory and cooperative framework for broad and effective measures to ensure that the Earth can be protected from major natural and human-created disasters. Such an effort should encourage and support cooperative actions that are—to borrow a phrase from the Outer Space Treaty—clearly "in the interest of all countries" and represent the "province of humankind".

Such aspirations are addressed later in Chap. 14 on a potential Global Sustainability Treaty.

Unfortunately, actual progress to achieve the ambitious U.N. goals remains far short of the stated aims. This can be seen by examining the progress—or lack thereof—against the nearly 200 targets that have been set to measure progress toward achieving the 17 goals. The sad truth is that realistically, when 2030 comes, the goals will remain unmet. This is because economic systems place a higher value on economic throughput, cheap energy, and maximizing short-term profits. Corporate activity that dominates worldwide economic production and services does not easily swallow the costs of coping with air, land, river, and ocean pollution. In brief, economic industrial objectives and environmental/sustainability goals still remain largely unaligned.

Private companies largely produce goods and services, and governments are relegated to the role of cleaning up the consequences or creating the vital infrastructure or regulations to keep economic engines running. The issues that must be addressed range widely. These concerns can be pollution, over-consumption of natural resources, transportation, national defense, disaster relief, weather forecasting, sanitation and water, labor issues, safety codes and inspections, health and education, and more.

Space System Monitoring and Enforcement

The world is changing dramatically with the worldwide introduction of broadband 5G cellular systems, which can accommodate everything from telephone calls to Internet streaming. The planned deployment of thousands of low and medium Earth orbit satellite constellations as well as high-powered conventional geosynchronous satellites will open up lower cost broadband communications to developing countries and rural and remote localities all over the world, including the oceans and Polar Regions. These high-speed digital communications will also support new applications such as driverless vehicles (i.e., cars, trucks, vans, and buses) and mobile video streaming services.

Yet, challenges to extending broadband connectivity to the underserved parts of the world will remain. This includes finding ways to supply electrical power to remote regions and achieving connectivity between localized Wi-Fi systems and lower cost satellites.

Other types of satellite systems can provide enormous new capabilities to support agriculture, forestry and forest management, water, river, and stream management, and to oversee fishing regulation, detect pollution, and assist with law enforcement. They can become enormously valuable when seeking to cope with natural disasters such as earthquakes, tsunamis that flood coastal areas, as well as other types of flooding events and volcanic eruptions. They can help assess areas of destruction, road blockages, power line damage, and much more. Satellite telecommunications, especially mobile satellite connections, can provide a lifeline between rescue workers and those handling logistics for relief supplies.

The challenge will be to connect these high-tech space capabilities with local industries as well as law enforcement and environmental protection agencies. As is almost always the case, the "last mile" of service connectivity is the biggest challenge of all.

The use of satellite imaging to monitor climate change and weather patterns is now vital. These enhanced space capabilities can detect increasing rates of lightning strikes, changing storm patterns, increases or reductions in rain rates, and changes in ocean and atmospheric temperatures. Satellite-based lightning trackers have for instance verified that there are now some 45% more lightning strikes occurring worldwide. This is a result highly consistent with climate change and the world's average temperature increase. Figure 10.1 shows global lightning strikes and their intensity, tracked through imaging from the GOES-T satellite.

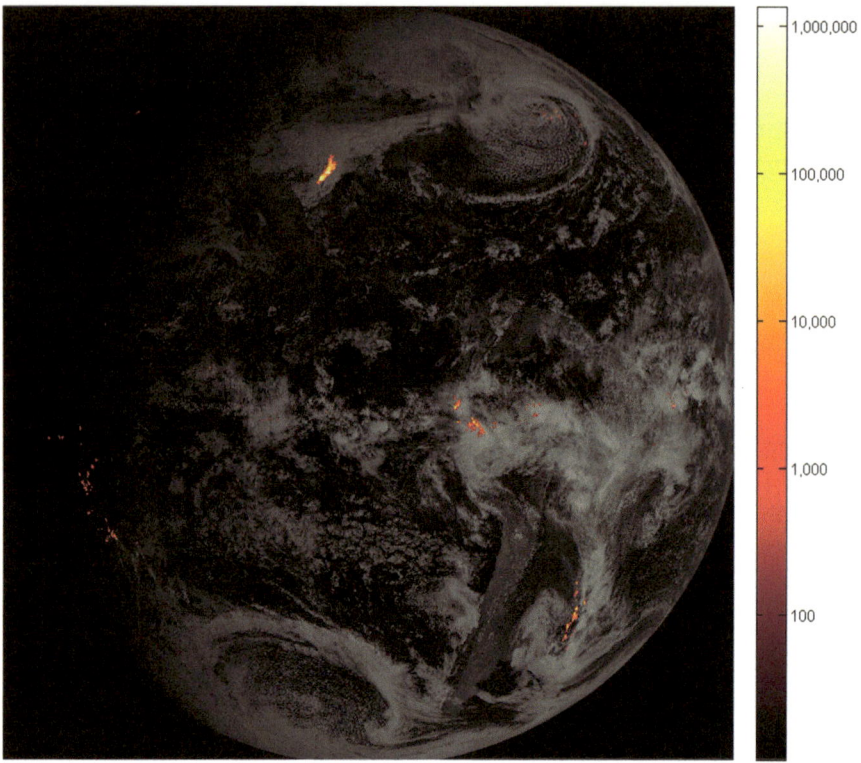

Fig. 10.1 The latest NOAA satellites are able to track lightning patterns. This image shows the intensity of storm and lightning in a global view. (Courtesy of NESDIS of NOAA)

The image in Fig. 10.2 reveals the desertification process in the Saharan Desert. The satellite tracking of vegetation in Africa throughout the course of a year seems to show a continent that is breathing in and out as plant life expands and retracts. But year after year, the desert regions slowly and relentlessly expand as the green areas shrink permanently [6].

The following range of candidates provides an initial menu of technologies, many of them space-based, that may help combat climate change and pollution in the near and longer term. They range from those requiring urgent support now to those requiring much more evaluation and more gradual implementation. The point is that there should be a globally coordinated process to accelerate the best and brightest concepts. The candidate programs are

Fig. 10.2 Map of Africa showing desert, semiarid, and tropical areas. (Courtesy of NASA)

not intended as a comprehensive list, but rather a starter list of new and often outside the box sustainability initiatives. The call within the Paris Accord to create a new technological framework is one of the key steps forward.

Many of the new technologies and systems in Table 10.1 are under active development, while others remain at a speculative stage. For those who wish to seek further information on some of these new systems, such as advanced solar cells [7], advanced space systems to monitor climate change in real time [8], solar power satellites [9], solar shields to protect Earth [10], and moving the Earth away from the Sun [11], the chapter endnotes are a place to start.

Table 10.1 Technical advances that might lessen the adverse effects of climate change

Nearer term	Middle term	Longer term
Developing automated data analytics to process remote sensing satellite data more rapidly, in order to address specific issues such as drying up of lakes, rivers, natural springs, and aquifers, and the monitoring of desertification rates, greenhouse gas release, etc. (possibly as public–private partnership arrangements)	Creating a solar power satellite system to beam down energy to the world 24 hours a day	Artificially creating a new ozone layer for Earth to protect against solar radiation, genetic mutation skin cancer, and damage to vital infrastructure
Development of new powering systems for vehicles that are electric, battery-powered, hydrogen-fueled, use advanced and lower cost fuel cells, or even use compressed air (these might be optimized for different nations)	Creating artificial refrigeration and insulation systems to prevent the Arctic peat fields from melting	Developing a solar shield against solar storms (i.e., coronal mass ejections) and creating asteroid protection capabilities
Developing advanced, lower cost, and more efficient solar energy/electrical energy conversion systems (i.e., using quantum dot technology or capturing wasted infrared radiation through hyperbolic dispersion. This might create narrowband radiation for conversion to electrical energy with up to 80% solar cell efficiency)	Developing a capping and treatment system to capture and convert large amounts of methane currently being released by emissions from vast mining operations in Siberia	Moving Earth's orbit outward to reach further from the Sun
Tax incentives for human birth control, as practiced in Singapore	Improved and lower cost birth control technology and with better and easier access	Chemical treatments to lighten the icecap albedo
Creation of a U.N.-sanctioned technology incubator or scientific council on new technologies that control climate change and global pollution	Iron enrichment and acidity reduction program. This would increase ocean bioproductivity and its function as a carbon sink	Global "heat pipes" to eject energy into space
Genetically altering cattle to reduce methane	Experimentation in darkening and colorization of clouds to increase reflectivity	

(continued)

Table 10.1 (continued)

Examples of possible new technologies that could transform global energy systems, help reduce climate change and pollution, and assist with planetary defense against cosmic hazards

Note: This chart was developed and copyrighted by J.N. Pelton with all rights reserved

Strategies for Sustainability

There is a need for both developed and developing countries to move ahead with all due speed to adopt new programs to cope with climate change and global pollution. These strategies, however, may be significantly different in different countries.

Convincing Developed Countries of the Urgency of Sustainable Programs

Countries with developed, industrialized economies may seek to buy their way out of difficulty and wall themselves off from those with developing economies. These approaches will only allow short-term relief. Addressing the challenge of climate change and global pollution must confront the cause. Pricing systems must adjust to reflect the real costs these crises.

The strategy for change has many dimensions. They include the following:

(i) Phasing in new technologies and green energy sources
(ii) Shifting employment out of industries that are wedded to carbon-based fuels and chemicals
(iii) Shifting service jobs to telecommuting operations and moving ideas more than people where possible
(iv) Governmental, economic, tax, and social reform programs designed to cope with the social and economic disruptions of super-automation and use such reforms in part to reduce pollution and adverse environmental impacts
(v) Emphasizing environmentally friendly educational programs, sustainability programs, and green lifestyles

There are two very significant problems in developed economies. First, most people in these societies are accustomed to a standard of living related to food, housing, heating and air-conditioning, and travel by personal cars and aircraft. They are generally reluctant to alter these luxuries of life to combat climate change. Second, business enterprises that are involved with carbon-based fuels, conventional energy, transportation, housing and buildings, among other fields, are heavily invested in infrastructure and systems that are driving climate change, and they are generally unwilling to accept that the "conventional way" of doing things must change and new technology substituted for the old.

In many ways, the problem is economic. The needed legal and regulatory changes are going to be resisted by both the general public and industry. The owners of oil, gas fields, and coal mines, refineries, car and truck manufacturing plants, of service stations, houses, apartment buildings, office building, shopping malls, and on and on, want to know about how to change without suffering huge economic losses.

Conclusions

Meeting the large challenges posed by hyperobjects, or what are described herein as long-term threats to humanity, seems today to be almost unsolvable challenges. The problems all seem simply too large in scope, too long in timescale, and too costly for an individual to try to take on by themselves.

Societal reforms are best achieved not only by prohibitions and bans but also by providing incentives to lead people, companies, and organizations toward desirable alternative behaviors. Tax breaks or other rewards for switching to electric cars, solar energy systems, and smaller, greener, smarter housing could help. If green practices become fashionable and less expensive, this allows an easier transition than a program that is forced on businesses or populaces through fines and penalties. The challenges of climate change are great, but logical steps forward exist. The conversion of non-binding goals in the Paris Accords to binding national commitments backed by enforceable laws would be a good next step. What is needed are established legal penalties for missing goals, and even more importantly, rewards for exceeding them.

The main idea presented in this chapter is that plans to address climate change as an issue unto itself is the wrong way forward. Only an integrated approach that looks at climate change in a big picture way will ultimately succeed. There is the old song about how all the bones in the body are interconnected. Likewise, all of the problems addressed in this book form the bones in the body of climate change.

References

1. Mufson, S., Dennis, B.: U.S. to leave Paris climate accord Nov. 4, but voters will decide for how long. Washington Post, 3 Nov 2020, p. A11
2. Sullivan, K.: Biden prioritizes climate crisis by naming John Kerry special envoy. CNN, 24 Nov 2020. https://www.cnn.com/2020/11/23/politics/john-kerry-biden-climate-envoy/index.html
3. Fuller, C.: As the U.S. threatens to scupper another all important climate treaty, we see why the Paris agreement might not collapse as easily as Kyoto. AOSIS, 2 Nov 2020. https://www.aosis.org/2020/11/02/lessons-from-the-kyoto-protocol
4. U.S. exits Paris, we are still in. https://www.wearestillin.com/paris-withdrawal-2020. Accessed 5 Nov 2020
5. https://sdgsinaction.com/. Accessed 20 Nov 2020
6. http://www.Earthobservatory.nasa.gov. Accessed 15 Nov 2020
7. Starr, M.: New device that channels heat into light could boost solar cell efficiency to 80%. Science Alert, 25 July 2019. https://www.sciencealert.com/device-that-channels-heat-into-light-could-boost-solar-efficiency-to-80-percent
8. Theia is changing humanity's relationship to the physical world. The Theia Space System. https://theiagroupinc.com/about-theia. Accessed 11 Nov 2020
9. First text of solar power satellite hardwar in orbit. Off Grid Energy Independence, 20 May 2020. https://www.offgridenergyindependence.com/articles/20709/first-test-of-solar-power-satellite-hardware-in-orbit
10. Pelton, J.N.: Our changing world and the mounting risk of a calamitous solar storm. Room Space J. (7) (2016) https://room.eu.com/article/our-changing-world-and-the-mounting-risk-of-a-calamitous-solar-storm
11. Brin, D.: Saving earth from an expanding sun. Room Space J. (7) (2016) https://room.eu.com/article/saving-earth-from-an-expanding-sun

11

Nuclear Waste and Pathogens

Nuclear waste is a heavy burden to lay on our children and their children and their children's children and their children's children's children and their children's children's, children's children.

–Rufina M. Laws

A disease outbreak anywhere is a risk everywhere.
–Dr. Tom Frieden, Former Director U.S. CDC

Introduction

Pollution was addressed in Chap. 4 and also to an extent again in Chap. 10, but largely in the context of negative environmental effects. Pathogens also were mentioned in Chap. 3, but mostly in the context of COVID-19 and pandemics.

There are a variety of pollutants, caustic materials, and pathogens caused by industrial and other commercial development activities, which can give rise to severe health problems, radiation hazards, and even death. These involve different concerns and regulatory actions. This is yet another area where various space systems and data analytics can help detect such problems, send out alerts, and support necessary recovery operations.

This chapter first considers the problem of nuclear waste and accidents that can threaten a community. It also outlines new types of satellite monitoring systems for reactor safety and ways of using space systems for the control of nuclear waste. In addition, it addresses the lack of appropriate safety practices associated with the operation of nuclear facilities and disposal practices for nuclear waste. This is followed by consideration of the medical or health problems that can arise from misuse or mishandling of pathogens, poisons, and other dangerous and caustic materials. Today, there are also serious concerns about protection against pathogens that might be brought back to Earth or transmitted from Earth through deep-space exploratory missions.

J. N. Pelton, *Space Systems and Sustainability*, https://doi.org/10.1007/978-3-030-75735-9_11

Historical Background

The history of human acts to despoil the planet with poisons, pollutants, dyes, and other substances is not a proud one. This has been accompanied by a long struggle to provide more governmental oversight of dangerous industrial and chemical sites, which remains lax to this day in many ways.

The first official international organization in the world, which is still functioning today, was formed to regulate navigation on the Rhine River and, somewhat as an afterthought, to also regulate the river's pollution problem. Before this commission was formed, printers located along the Rhine routinely drained noxious inks and dyes into the river. The process to bring some regulatory control to the river began on October 15, 1804, with the creation of the Rhine River Commission. This set up a uniform toll for the Rhine River and an administration for it at Mainz, Germany, establishing the principle of free navigation for all who paid the toll.

These concepts and more were internationally formalized in Appendix 16B to the final document agreed at the Congress of Vienna held in 1815. This event created the Central Commission for Navigation of the Rhine, which is now headquartered in Strasbourg, France. The mission of this Central Commission was to establish or further standardize navigation procedures along the Rhine, moderate tolls, and fine polluters. The Rhine River, which flows through a half dozen European countries, was deemed an international resource that finally had international controls applied to it. So the world's first international organization was born, with the control of pollution a part of its mission [1].

The Congress of Vienna was a pivotal point in the history of international regulatory processes. Held at the end of the Napoleonic Wars, it began a new era of international cooperation among the countries of Europe. Since that time, and especially post-World War II, there has been significant progress in the creation of international law, binding treaties and conventions, and many other types of international organizations.

Today, there is the overarching United Nations Organization and myriad specialized agencies of the United Nations that support its overall mission. Many of these are important to the control of nuclear and radioactive materials, contaminated food, problems with radiation and radiation sickness, and global disease and pollution. They include:

(i) *The World Health Organization (WHO)*, whose mission includes establishing international standards for health, radiation hazards, and control of pathogens, as well as providing global alerts for pandemics [2]

(ii) *The Food and Agricultural Organization (FAO)* that sets standards for food safety and related controls of pollutants, fertilizers, and insecticides [3]

(iii) *The International Atomic Energy Agency (IAEA)* to control and regulate nuclear and radiological materials [4]

(iv) Several agencies to address atmospheric pollution, global ecology, climate change, environmental concerns, and noxious materials. These include the *World Meteorological Organization (WMO)* [5] and the *U.N. Environmental Programme (UNEP)* [6]

(v) *The U.N. Office of Disarmament Affairs (UNODA)* to provide help with new procedures involving the safe shipment of nuclear and radioactive materials [7]

The True Costs of Nuclear Power

The idea that nuclear energy and power plants could be a low-cost, safe, and sustainable source of energy was advocated by the "Atoms for Peace" initiative in the 1950s some 70 years ago. The environmental premises on which this technology was originally based included the following:

(i) Nuclear waste can be safely disposed of without contaminating water supplies of local populations.

(ii) The cooling systems for nuclear power plants will not create serious thermal pollution.

(iii) These systems will always be designed with fail-safe capabilities against earthquakes, tsunamis, and other natural disasters.

(iv) These systems at end of life can be shut down and phased out of operation in a safe and secure manner.

(v) The miracle of atomic reactions will provide renewable, clean power at the lowest cost.

The devastating events that have occurred at Three Mile Island, Chernobyl, and the Fukushima Daiichi nuclear sites should provide convincing evidence that all these reassurances were flawed. The huge, mounting costs associated with radioactive waste disposal as well as the 16 radioactive "Superfund" sites across the United States also belie claims that nuclear energy is altogether cheap and safe. On top of these concerns, the constant fear of nuclear power plants being possible targets of attack by techno-terrorists further compounds the risk that these facilities pose.

Although environmentally conscientious public figures such as Bill Gates have expressed optimism in new types of nuclear plants and cooling systems (i.e., nuclear plants that no longer use water as the primary coolant), the historical evidence is against the premise that nuclear plants based on nuclear fission technology can be safe [8].

Research efforts today are rightly focused on reducing the costs of solar, wind, hydroelectric, ocean wave and ocean thermal energy conversion, geothermal, and other sustainable electrical power systems. Much of these research projects should focus on extending the lifetimes of these systems, which will help prove their longer-term economic validity.

Often in discussions of nuclear or carbon-based fuels and their overall cost, the cleanup, environmental costs, and/or health risks are not included in the calculations. These considerations can double or even triple the net cost of the energy charges to consumers. Further, the cost over time must also be considered. Wind, geothermal, solar, tidal, or ocean thermal energy conversion are energy sources that keep on giving, millennia after millennia.

Miscalculated Risks and Improper Storage

Nuclear power plants are designed to withstand conditions that are determined to be likely for the areas in which they are built. With disturbing consequences, these determinations have more than once proven incorrect.

The Indian Point Nuclear Reactor Center with its three reactors is located on the Hudson River about 30 miles (48 km) north of New York City. This was thought to be in a very safe seismic area when it was built. Subsequently, experts detected fault lines that increased safety fears. Safety officials now rate this nuclear power center as the "at most risk" facility in the United States. It is scheduled to close in 2021 or 2022 due to lack of revenue production and elevated safety risk [9].

The Diablo Canyon Nuclear Power Plant between Los Angeles and San Francisco, California, was built to withstand a 6.75-magnitude earthquake and has now been upgraded to withstand a 7.5-level event. This facility was significantly upgraded when it was discovered in 2008 that one fault line in particular was just 2000 feet (650 m) from the reactor and just 985 feet (320 m) from the crucial water intake system. There have been organizations such as the Union of Concerned Scientists that have pressed the Nuclear Regulatory Commission (NRC) to close down this facility due to safety concerns, but nearly 15 years after the fault line was discovered, it continues to function [10].

The experiences in Japan with the Fukushima Daiichi site alone should be convincing. The Fukushima Daiichi partial meltdown stemmed from a miscalculation of the earthquake threat, which created the very powerful tsunami that destroyed the three reactors at that site. In 2016, the Japanese government originally estimated the costs of this disaster at over $200 billion (US). But the Japanese Center for Economic Research, a private research foundation, later provided perhaps a more realistic estimate that ranged from $470 billion to $660 billion. As of April 2019, *CleanTechnica* news provided a comprehensive estimate of the direct and indirect costs of accident at close to $1 trillion [11].

The *CleanTechnica* estimate does not include the distress to families whose loved ones were exposed to the high levels of radiation released in the accident. Figure 11.1 shows the ill effects that may come to anyone exposed to a high number RADs (radiation-absorbed dose, measured also in the unit "Grays"). The noxious effects to the brain, skin, lungs, vital glands, bone marrow, and GI tract are quite significant and can be lethal. Death may come within a day if the radiation levels are sufficiently high.

There are major problems occurring with improper nuclear waste storage all over the world. Experts estimate that nuclear waste now exceeds a quarter million metric tons of highly radioactive material. This includes thousands of tons of waste from nuclear power reactors as well as perhaps millions of liters of radioactive liquid materials that are leftovers from nuclear weapons production sites. Most of this waste sits in storage areas near both nuclear power plants and weapons production facilities worldwide. About 35% of this is stored in the United States, and the other 65% is distributed around the world in some 50 other countries [12].

The struggle seems to be how to manage near-term health problems and possible contamination against long-term risks for unsuspecting generations many years from now. There is no clear, healthy answer to what can be done with this massive amount of radioactive material. Proposals to reprocess the waste are met with convincing answers that this would be much too expensive, would create new safety and health risks, and would create opportunities for terrorists to acquire nuclear materials for "dirty bombs." Such recycling would ultimately be able to cope with only a portion of the waste materials.

As politicians and regulators debate what to do, the problem continues to grow. Current storage facilities are far from secure. Instances of the concrete

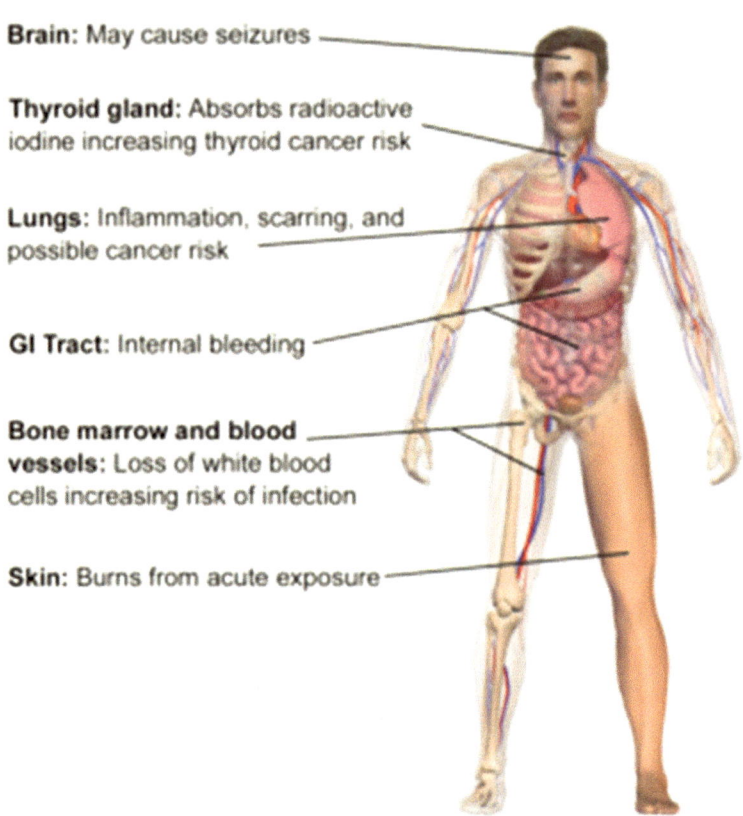

Brain: May cause seizures

Thyroid gland: Absorbs radioactive iodine increasing thyroid cancer risk

Lungs: Inflammation, scarring, and possible cancer risk

GI Tract: Internal bleeding

Bone marrow and blood vessels: Loss of white blood cells increasing risk of infection

Skin: Burns from acute exposure

Selected Risks from Radiation Sickness

Fig. 11.1 Radiation risks from high levels of X-ray and UV exposure. (Courtesy of Wikimedia Commons)

and steel waste storage containers being jostled about and sometimes thrown into piles by earthquake events heighten existing concerns. Efforts to find locations in seismically stable areas for storage, perhaps within mountains, ultimately find objections from the public. Howard Hughes himself managed to prevent planned storage sites in Nevada.

The underground tanks for storage of nuclear weapon byproducts were constructed beginning in the 1940s in Hanford, Washington (Fig. 11.2).

It is claimed that at the Hanford, Washington site, which was used for decades to create nuclear weapons, there is something like 200 million liters of accumulated waste stored in underground tanks. This liquid radioactive waste is still waiting to be processed. It is alleged that the "safe lifetime" for these tanks, now well over 70 years old, has been exceeded and that an

Fig. 11.2   Picture of large underground storage tanks under construction in the 1940s, built to hold radioactive liquids at the Hanford site in Washington, United States. (Courtesy of the U.S. Department of Energy)

estimated one-third might be leaking radioactive liquid deep into the soil. There is fear, expressed by the Union of Concerned Scientists, that this potential leak could eventually reach the Columbia River and lead to severe pollution and public health problems [12].

And this is but one site in one country. Currently, there is a multibillion-dollar facility near completion to convert the radioactive liquid into a glasslike substance that can be stored more safely. This process is known as *vitrification* or *vytrification*. The problem is that there is a huge amount of radioactive waste that remains to be processed beyond that which might be handled by a few vitrification processing plants [13].

There are now new types of nuclear power reactors that can use the spent fuel without "recycling" it. Nevertheless, questions remain. Even if a better way is found to process this mass amount of nuclear waste, it should be clear by now that this is not the answer to the electrical power needs of future generations. Is it not better to find a way to make recyclable energy sources such as solar and wind energy systems the priority?

Cyber Attacks Against Nuclear Facilities

The good news about the massive cyberattack against the United States during 2020 is that it did not seem to include attacks on the computer networks associated with the controls and safety mechanisms for nuclear power plants. Nevertheless, one of the private software suppliers that was a direct target of the attack was the Solar Winds Company and its Orion software, which had previously been used at a number of nuclear sites up until 2011. Cyber security is today a very serious consideration that will be addressed further in Chap. 13.

In the areas of radioactive pollution, radioactive waste storage, and cyber-threat protective systems, it will be increasingly possible for the newest space and related IT and AI technologies to create monitoring, detection, warning, and emergency response systems. It would be possible to install IoT radiation monitors at the sites of all nuclear reactors as well as nuclear waste storage plants in order to form a global alert system. It could use Internet of Things (IoT) or possibly SCADA technology on the ground as the prime monitors. Robotic systems could be equipped with IoT sensors to reduce exposure to radiation. The use of robotic systems could also be applied to the detection of pathogens in the case of high security medical research laboratories.

Pathogens

Current industrial processes, biological research activities, overuse of antibiotics, and other human activities can give rise to dangerous pathogens. These health risks can in some cases be highly localized, but in other cases, a released pathogen might ultimately create a global pandemic, as addressed in Chap. 3. As shown with nuclear waste and practices, a lack of foresight about safety practices and containment issues can lead to billions of dollars if not trillions of dollars in societal damages and widespread health concerns.

Radioactive materials and nuclear may be creating radiation sickness today and genetic damage to future generations. Likewise, the current use of antibiotics against today's viruses and bacteria may be endangering future generations in the decades to come. Today, antibiotics are widely used to treat various diseases, whether they are truly effective against those diseases or not. This can make the antibiotics less effective against certain now-resistant bacteria and viruses. Such overuse is apparently giving rise to so-called "superbugs."

The pathogen viruses able to attack the human body and bring on deadly virulence are currently numbered at 1400. Yet this is just a small subset of the microorganisms living on Earth. In just a teaspoon of soil, one can find about a billion live microorganisms. The number of viruses that exist on Earth and affect animals, plants, and other organisms have been estimated to be a truly huge number, expressed as 10^{31}. The very good news is that most of these microorganisms are not dangerous, or at least do not bring so-called virulence to the human body [14].

The pathogens that do attack the human body can bring great suffering and death in the form of pandemics (see Chap. 3). The current worry with the Covid-19 is that novel coronaviruses attack the body in an unfamiliar way, and the body cannot develop antibodies that fight back effectively. Many variant versions of this virus are spreading across the world, and there is some doubt about the efficacy of the vaccine against all of the new strains. Other coronaviruses can be expected to evolve.

There are now also increased concerns about the potential destructive powers of dormant viruses. For instance, a "viable virus" has been extracted from a piece of amber that was estimated to be at least 34,000 years old, and possibly was well over 100,000 years old. Such ancient viruses, for which humans have no natural defense, might prove as destructive as a virus brought back from outer space.

Particularly because of untrue theories about COVID-19, there are emerging concerns about highly infectious pathogens that could come from such sources as research laboratories at universities, high containment labs (HCL) of governmental facilities, or industrial sites that are developing new drugs or organic products and applications.

Space and Informatics Technologies

The dangers of the various byproducts from nuclear power plants, industrial processing plants, polluted streams and waterways, etc. are today a clear area of global health concerns. So too is containment of pathogens, poisons, and other caustic materials. The key question is how to use the best and brightest of current space and ground sensor technologies to contain such threats.

There are today literally billions of smart, compact IoT systems that can be integrated with quite small satellite antenna systems. These tiny sensors and satellite antennas can detect dangers and send instant alerts with just a few bytes of information from a smart IoT sensor embedded in a flat phased-array antenna located virtually anywhere on the ground, on the oceans, or even flying in the sky [15]. Such units could send out an alert if a sealed door is opened illegally, a new source of radioactivity is sensed in an unsecured area, or a dangerous pathogen is detected in an uncontained area of a medical lab.

Companies such as Intelsat, Globalstar, the Telesat LEO constellation, and perhaps a dozen other systems that can provide Automatic Identification Services (AIS) are also now equipped to receive data message signals from embeddable satellite transmitters. This capability can allow developers to rapidly design, develop, and even manufacture their own sensors and digital alert system. Integrated chipsets into satellite antenna products are today used for motor vehicles, operational equipment of various types, environmental and pollution monitoring, wildlife tracking, agricultural remote data monitoring and reporting systems, and more. Global coverage well beyond the reach of traditional cellular service areas is now possible using these satellite terminals, which can be as small as a credit card (Fig. 11.3).

Fig. 11.3 The SX100 flat antenna with embedded technology, compatible with the Globalstar Satellite Network. It is the size of a credit card. (Credit: Globalstar)

These types of mobile capabilities are now being inserted in consumer products from automotive products to aircraft. The design of systems to alert the world to a problem of leaking radioactive materials, dangerous pathogens, caustic materials, or poisons should be deployed around the world. In many cases, these sensors and detection monitors can be simply linked to terrestrial IT networks, but in many where transport is involved or there might be a large number of areas or points to be covered, satellite systems will be the more effective mode of connection. Too often hazardous facilities are monitored by the operator of dangerous sites. Experience has taught the world. It is better to separate the regulatory oversight authority from those carrying out dangerous and potentially unhealthy or even potentially lethal activities if safety is not strictly observed.

Conclusions

Today there are many concerns about things that might compromise human health or natural life. These can come from dangerous byproducts of nuclear-related activities, radioactive waste materials, or various types of new pathogens.

There is need of a comprehensive and integrated global monitoring and alert system that could be instantly triggered if there were a radioactive leak or serious problem with nuclear waste, or any one of the many threats that can happen in a complex world. Such a system could be built on the back of existing capabilities already developed by the WHO, the FAO, perhaps the IAEA, or even through the UNODA. But today, these separate systems exist as islands of competencies and are not designed to easily to work together.

References

1. The Central Commission of the Rhine River (CCR), the Organization. https://www.ccr-zkr.org/11010100-en.html. Accessed 16 Nov 2020
2. World Health Organization. https://www.who.int/. Accessed Nov 2020
3. Food and Agricultural Organization (FAO)
4. Global Nuclear Safety and Security Network. https://www.iaea.org/services/networks/global-nuclear-safety-and-security-network. Accessed 10 Jan 2021
5. World Meteorological Organization. Environment. https://public.wmo.int/en/our-mandate/focus-areas/environment. Accessed 10 Jan 2021

6. UNEP. Time is running out for coral reefs: new report. https://www.unep.org/news-and-stories/story/time-running-out-coral-reefs-new-report. Accessed 10 Jan 2021

7. U.N. Office of Disarmament Affairs. https://www.un.org/disarmament/. Accessed 10 Jan 2021

8. Gardner, T.: Bill Gates' nuclear venture plans reactor to complement solar, wind power boom. Reuters, 27 Aug 2020. https://www.reuters.com/article/us-usa-nuclearpower-terrapower/bill-gates

9. Becker, R.: New York City's closest nuclear power plant will close in five years. The Verge. 6 Jan 2017. https://www.theverge.com/2017/1/6/14196006/new-york-state-indian-point-nuclear-power-plant-close-2021

10. Union of concerned scientists, 3 Sept 2013. https://www.ucsusa.org/resources/diablo-canyon-and-earthquake-risk

11. Fukushima's final costs will approach trillion dollars, 2019/04/16, cleantechnica.com. Accessed 11 Jan 2021

12. Jacoby, M.: Association of concerned scientists "as nuclear waste piles up, scientists seek the best long-term storage solutions". C&N 98. (12) (Mar 2020)

13. Using glass for nuclear vitrification. https://www.azom.com/article.aspx?ArticleID=18307. Accessed 11 Jan 2021

14. Microbiology by numbers. https://www.nature.com/articles/nrmicro2644. Accessed 10 Jan 2021

15. Pelton, J.N. (ed.): Handbook of Small Satellites: Technology, Design, Manufacture, Applications, Economics, and Regulation, pp. 511–598. Springer Press, Cham (2020)

12

Natural Disasters

Preparation through education is less costly than learning through tragedy.
–Max Mayfield, Director, National Hurricane Center

It wasn't raining when Noah built the ark.
–Howard Ruff

Introduction

In the ancient world, natural disasters were perceived as preordained events, perhaps as punishment from the gods. In the modern world, there is a much better understanding of the many forces that shape these phenomena, along with a range of technologies that can help mitigate their impact and quicken recovery.

The natural course of disasters in terms of frequency of occurrence and number of deaths follows a clear historical pattern. For many decades, the preponderance of disasters has been hydrometeorological and climate-related, including hurricanes, cyclones, tropical storms, tornadoes, tidal waves, and flooding. Next in frequency and devastation are geologic events, such as earthquakes, tsunamis, landslides, volcanic eruptions, etc. Finally, there are natural biological catastrophes, such as pandemics and major outbreaks of sickness.

This chapter reviews natural disasters of all types, the frequency of their occurrence, their relative dangers, and what more can be done to mitigate such events. It also considers these disasters against major trends of change, including urbanization, population growth, and infrastructure dependence.

Disaster Preparedness

A disaster is not simply something to respond to after the fact. A great deal of training and certification, resource preparation, communications, and

J. N. Pelton, *Space Systems and Sustainability*, https://doi.org/10.1007/978-3-030-75735-9_12

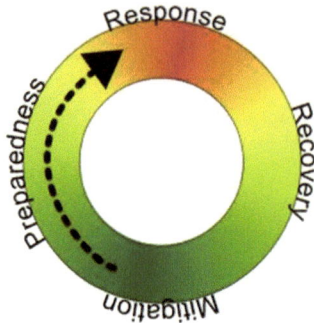

Fig. 12.1 The disaster management cycle: preparedness, response, recovery, and mitigation. (Graphic courtesy of the U.S. Federal Emergency Management Agency)

transportation capabilities must be ready and in many cases pre-deployed. The continuous cycle of activity associated with disaster preparedness is shown in Fig. 12.1.

There are many stories about disaster relief workers who are not properly trained, and disaster recovery breakdowns are legion. For instance, many personnel have not been trained in the use of satellite phones or other specialized relief equipment. There are also almost laughable problems with relief workers and critical personnel who are not able to receive the right credentials in time to provide rapid relief where it is needed. There are also sometimes problems with police, fire, medical, or army personnel using a range of different codes that sometimes mean the opposite thing when relayed from one group to another.

Of the four aspects of disaster management, it is truly the preparation stage that can save the most lives and limit financial losses. Something as basic and wide-reaching as organizing a mass evacuation against an oncoming storm is often the best strategy.

The Rising Levels of Natural Disasters

The period from 2000 to 2019 has been a time of increasing crises as measured by almost any dimension. This might be in terms of violence of rain, storm, flooding, and meteorological events, or increasing levels and frequency of lightning strikes and magnitude of property loss. During this two-decade-long period, the EM-DAT database has recorded a total of 7348 disaster events around the world. This resulted in the loss of 1.23 million lives, or an average of about 60,000 per year. These events also affected nearly 4 billion people in some form.

If one compares the last 20 years to the 20 years prior (i.e., 1980–1999), the statistical increase is disturbingly clear. The recorded disasters for the earlier two decades were much fewer, at a total of 4212. Thus, disaster events increased by a total of 3136 for the 2000–2019 period in an increase of about 75%. The total property losses rose from $1.63 trillion (during 1980–1999) to $2.97 trillion dollars (during 2000–2019), demonstrating a significant increase even allowing for inflation.

Fortunately, when the increase in global population during the two periods is taken into account, the loss of life on a per capita basis showed improvement. There were a total of 1.19 million in disaster-related deaths for the 1980–1999 period. This compares to 1.23 million disaster-related deaths from 2000 to 2019. This would indicate a slight reduction on a per capita basis [1].

A similar statistical profile can be seen for the same period covering just the United States. Figure 12.2 breaks down four types of natural disasters from 1980 to 2018 in the United States [2]. This chart shows that storms and meteorological events as shown in green are the most deadly."

There are certain criteria for entering a disaster event into the U.S. Agency for International Development (USAID) Office of Foreign Disaster Assistance/ Centre for Research on the Epidemiology of Disasters (OFDA/CRED) database, known as EM-DAT. These are:

(i) At least 10 people killed
(ii) At least 100 people adversely affected
(iii) A formal declaration of an emergency or a formal request for international assistance

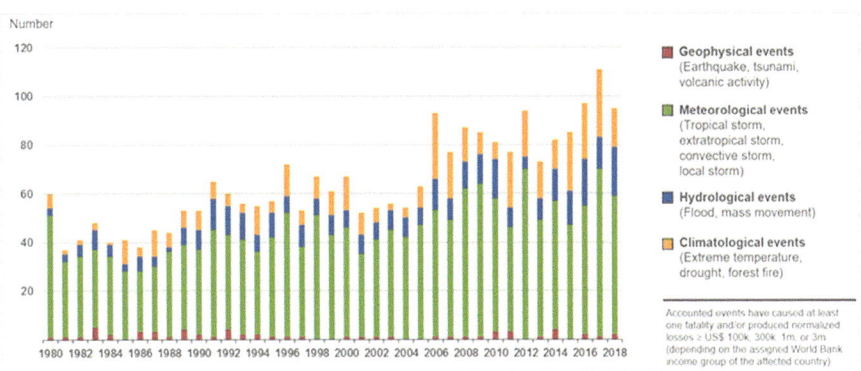

Fig. 12.2 Natural disasters from 1980 to 2018. (Courtesy of OFDA/CRED EM-DAT Database)

Key Data for Disaster Relief Databases

- Boundaries: Local, regional, national, international
- Chemical/manufacturing chemical facilities, toxic release sites, Hazardous materials storage/nuclear storage facilities
- Education: schools and libraries
- Emergency services: fire, police, EMS, public works, seats of Government, Continuity of Operations sites, prisons and other Secure facilities
- Energy: gas storage and processing, power plants, pumping stations
- National symbols: public venues
- Tourism and cultural heritage (archaeology)
- Public health: hospitals, pharmacies, blood banks, vet clinics, Morgues, and mortuaries
- Potential shelter sites, mass feeding sites, decontamination sites,
- Special areas (military and prisons)
- Weather: sirens, weather service sites, monitoring sites
- Dams and other potential hazards; flood zones

Fig. 12.3 The rising level of damages as a result of global disasters. (Courtesy of EM-DAT, OFDA/CRED)

Disasters are numbered by an eight-digit citation, with the first four numbers indicating the year and the next four indicating its sequence in that year, starting with 0001 [3].

Figure 12.3 shows one index that demonstrates rising damage levels, which derives from the EM-DAT database. This data is developed by the U.S. AID Office of Foreign Disaster Assistance Office in conjunction with the Centre for Research on Epidemiology of Disasters (OFDA/CRED). The advent of disasters is not consistent from year to year. Clearly, mega-events such as the Kobe earthquake and Hurricane Katrina distort the reporting for any 1 year. In addition, part of the rise might be attributed to better collection and reporting of data. Still, cumulative data from decade to decade clearly shows a clear upward trend.

Remote sensing and meteorological satellites are today central to understanding how, where, and how fast things are changing. Space systems can show much more precisely the rate of change, the precise location where these changes are occurring, and much more. Such systems, plus GNSS space navigation and global telecommunications satellites, can also provide key support for preparation, response, recovery, and mitigation in a systematic way—and do so for all over the world. A number of programs have developed to make space systems more effective for disaster forecasting and relief efforts around the world. The following section highlights the various ways that this is now happening.

New Space Systems to Cope with Natural Disasters

Today's space systems can detect major disaster threats, provide timely warnings to affected populations, direct recovery traffic, and help with mitigation activities. The problem is that not all countries and response and recovery teams are well trained in these space-based tools. A key objective for the future is therefore an improved interface between response and recovery teams, particularly in developing countries, and those operating space-based systems. As space-based recovery expert Prof. Scott Madry has said:

> Disaster response and recovery in today's world brings together two extremes. On one hand there are very high tech space technologists and on the other there is the very down-to-earth and practical world of recovery teams who trust techniques they have been trained to use for years. Sometimes they mix like oil and water. Trust, open communications, and mutual understanding of strengths and talents are crucial to forming effective rescue and recovery teams that rely on each other talents and knowledge [4].

Some of the most important existing capabilities are in the area of satellite remote sensing systems. These include:

(i) More rapid updates (i.e., increased temporal resolution by means of a number of large-scale satellite constellations in LEO).

(ii) Hyper-spectral resolution that allows a much better understanding of the changes that are occurring in different climates. Earlier systems divided remote sensing into a small handful of bands, whereas today's hyper-spectral remote sensing satellites might divide infrared, optical, and ultraviolet bands into as many as a hundred spectral bands.

(iii) New radar satellite imaging that can peer into cloud-covered areas and a new type of tool to help examine structural damage to buildings, dams, bridges, or other problems caused by high winds, floods, or earthquakes [5].

Meteorological satellites help predict weather patterns and track the speed, intensity, and direction of storms. The latest technical innovations allow them to track lightning strikes that can reveal intensification of storms, along with vectors of movement. There are other capabilities that include broadband and mobile communications, navigation and positioning services, and data acquisition from buoys, ground- and sea-based monitors, and much more.

The speed of analysis through the use of supercomputers, or employing the computing power of hundreds of volunteers, can provide near real-time information about impact disaster areas. Intelligent heuristics and mapping techniques, when combined with remote sensing information from satellites, high-altitude platforms even helicopters, and quick data processing, can prove crucial to response and recovery teams.

One example is the use of AI heuristics and mapping techniques for rapid analysis after two volcanic eruptions occurred from Mount Agung in Bali in 2017 and 2018. This process was able to clearly identify the impacted areas where ash had spread. It distinguished the materials from the new eruptions versus the earlier eruptions in 1963–1964 [6].

In the best circumstances, there is a database for all areas of the world where a major disaster might occur, with critical information already developed and available by instant data network connection. The Red Cross and Red Crescent network has created critical data layers for the four phases of the disaster cycle. The system used by the network is shown in Fig. 12.4.

Most systematic data collectors focused on future disaster recovery will use a similar approach. The objective is to collect critical information and store it in geographic information systems (GIS) so that it is available by internet access well before a disaster strikes. Geographic information systems (GIS) are a special way to store information related to a specific location on planet Earth. A GIS can store detailed information for a specific location such as a house, a playground, a store, or a service station. In the latest GIS database, there are also ways to store information for something like a 30-story apartment building, showing information on a floor-by-floor layout that a fireman, policeman, rescue worker, or first responder might need to know. This allows key infrastructure information about individual homes, buildings, streets, dams, utilities, power stations, IT network facilities, police and fire stations, water reservoirs, and more to be stored in GIS format.

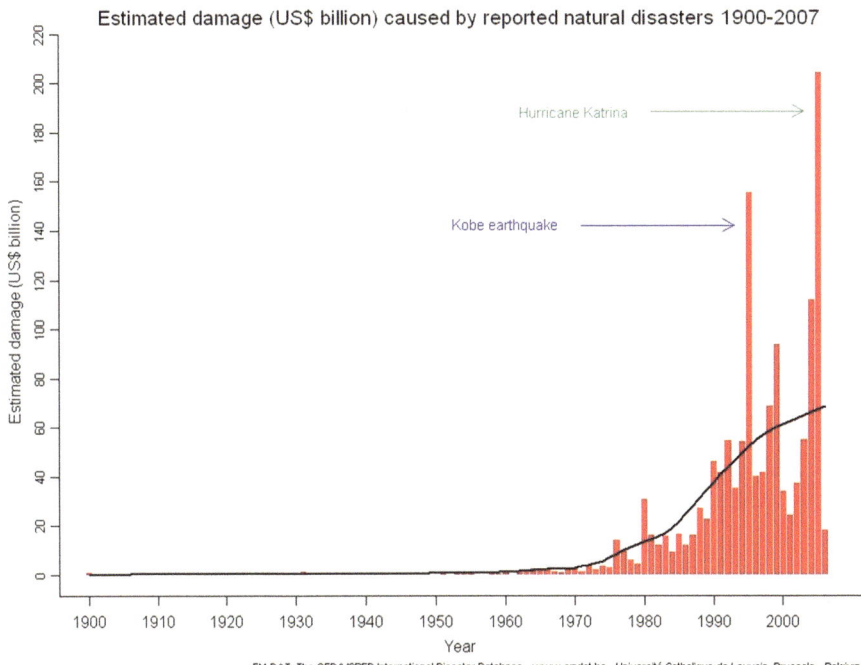

Fig. 12.4   Key data sought for GIS disaster preparation databases. (Source: Red Cross International)

Key Institutions, Resources, and Disaster Centers

There are a growing number of ways that national and global coordination for disaster relief have grown in recent years.

GeoCONOPS

The Homeland Security Geospatial Concept-of-Operations (GeoCONOPS) is a blueprint that identifies best practices, technical issues, and data sources for effective geospatial information and tools, all to support incident management in the United States. It has been recognized that there is a general lack of standardized processes for geospatial data management in major disasters. This is monitored and guided by the Geospatial Interagency Oversight Team (GIOT).

OFDA-CRED EM-DAT Database

One helpful way to prepare and respond to disasters is to know where and when they have occurred, along with their relative severity in terms of loss of life and injuries, property loss, duration, etc. The OFDA-CRED EM-DAT database, which has been mentioned already, is a long-running source of information for such data. In addition, it provides useful analytical tools to interpret the data.

Asian Disaster Preparedness Center (ADPC)

As a consequence of the great Kobe earthquake, the Japanese government funded the creation of the Asian Disaster Preparedness Center in Kobe, Japan. It is important to note that this center includes preparedness in its name. The center has, for instance, placed satellite terminals at all post offices across Japan in the hopes of sustaining broadband communications at these locations even in the event of another major earthquake.

In addition, the center provides training and disaster preparedness courses, along with awareness campaigns, throughout the entire Asian area, and is now doing so in partnership with the Bill and Melinda Gates Foundation. ADPC is also providing strategic and some financial support for the National Public Health Emergency Operations Centers throughout the region.

In addition, the ADPC is developing an online repository with regard to disaster preparedness. It has created a program for countries of the "Global South" to exchange information in what is called a south-south learning and knowledge exchange. Lastly, in partnership with USAID and NASA, it has developed the SERVIR-Mekong program, which uses geospatial data to help address climate change challenges among Lower Mekong countries [7].

The center's relationship with the Pacific Disaster Center (to be discussed next) has focused on commonality of formatting and insertion of English with regard to key information and data layers. This allows the two centers to share information more easily. The two have also shared satellite remote sensing data in the case of emergencies affecting Asia or the greater Asia-Pacific Rim.

Pacific Disaster Center

This is a facility established and operated by the U.S. government in Hawaii. It was authorized in 1992 after Hurricane Iniki devastated the island of Kauai, and it became operational in 1996. This center now offers constant online updates on disasters in the greater Pacific Rim countries and a current status report. It provides information through a dashboard display and also manages and operates a "DisasterAware" program. There is also a Disaster Alert system, which currently reaches over two million subscribers.

The center's sanctioned activities include:

1. Provide early warning alerts & technologies
2. Provide "risk and vulnerability assessments" (RVAs)
3. Conduct international working partnerships related to disaster warning
4. Working with developing countries to improve disaster preparedness and recovery, both through training and technical assistance
5. Provide humanitarian assistance and carry out disaster relief operations as requested [8].

Agreements for Remote Sensing and Communications Satellites

The Tampere Convention on the Provision of Telecommunication Resources for Disaster Mitigation and Relief Operations (Known Simply as Tampere Convention)

This convention began as a result of blockage at national borders when importing emergency portable ground stations for disaster relief needs. This blockage might sometimes last for days and might be accompanied by demands for high tariffs, even though the equipment was being donated and was intended to be in the country in question for days or weeks before being deployed for another emergency use. The need for this type of international agreement was first identified at an Annenberg Conference in the early 1990s and finally came into force in 2005. It has been ratified by some 50 countries to date.

The Tampere Conventions allow satellite communications equipment required for emergency communication to be shipped across borders without frequency allocation requirements, customs duties, importation licenses, or other requirements that might usually cause delays with customs officials. It also calls on members to promote the use of telecommunications in disaster response. In addition, it removes any restrictions on the entry or free movement of relief teams while operating in relation to the national disaster for which the Convention was invoked.

While this treaty is formally in effect, it is important to notify top homeland security, customs, and other officials, since the Convention is not widely known and delays have occurred even for countries that have requested the equipment under the Tampere Convention. The development of satellite telephones, such as those operated with the Iridium, Globalstar, Inmarsat, and Thuriya satellite systems, has lessened the need for emergency trans-border shipments. Further, mobile earth stations that can operate with the new LEO satellite constellations such as Starlink and OneWeb will further reduce this need. This is because the handheld satellite phones and the stations needed for emergency will largely already be in the country in question [9].

The International Charter on Space and Disasters (Known as the Disaster Treaty)

Another agreement, known as the International Charter on Space and Disasters, was agreed in 1999. This agreement provided a process to access to the most up-to-date satellite remote sensing data, and to get international assistance to process it so that it can be most helpful to relief crews. This process now works through space agencies and, in the case of the United States, by means of national meteorological agencies operating meteorological satellites. The idea was to make useful satellite sensing data available in an almost automatic way in case of major disasters, and to do so in a rapid and globally coordinated manner.

The initiative to create this charter originated with the French Space Agency (CNES) and the European Space Agency starting in 1999. Canada signed in 2000, and other countries quickly followed. The signatories now include the Argentine Space Agency, the U.S. Geological Survey, the U.S. National Oceanic and Atmospheric Administration (NOAA), the Japanese Aerospace Exploration Agency (JAXA), the China National Space Administration, the German Space Agency (DLR), the Korean Space Agency (KARI), the European Meteorological Satellite Organization (Eumetsat), and the Brazilian Space Agency (IMPE).

There is a formal process where national contact organizations specified under the Disaster Treaty Charter request activation of this international agreement. As of early January 2021, the Disaster Treaty Charter has been activated in response to over 300 major disaster requests. Over 120 countries have participated in its use. Each of the signatories has committed both satellite imaging data and analysis in support of disaster relief requests.

In 2010, the charter was amended to allow for what can best be described as open, universal access. Now, any national disaster management authority in the world can submit requests for activation of the charter. Thus, the country making the request for assistance does not need to be an actual charter member [10].

United Nations Resolutions and Mechanisms

The United Nations' General Assembly unanimously adopted in 1991 a resolution spelling out its global role with regard to natural disaster response and appeals for relief. This was known as resolution 46/182. The resolution specifies how the U.N. would support humanitarian needs in the event of such a disaster, including housing, emergency aid for the homeless, and so on. Any member-state of the United Nations can call upon the U.N. Central Emergency Response Fund (CERF) through the U.N. Emergency Relief Coordinator.

There is also an Inter-Agency Standing Committee (IASC) for emergency relief that coordinates specific national needs, whether they be for communications, food, housing, medical care, clothing and blankets, etc. The resources of many specialized U.N. agencies, such as the World Health Organization, the Food and Agricultural Organization, the International Bank for Reconstruction and Development, the International Telecommunication Union, Engineers without Borders, the Humanitarian Coalition, and others, are a part of this coordinative process [11].

In 2006, the United Nations Office of Outer Space Affairs developed a program known as the United Nations Platform for Space-Based Information for Disaster Management and Emergency Response (UN-SPIDER). The objective of this platform was to develop relevant information and emergency response solutions largely targeted to meet the needs of developing countries. This addressed both preparations and effective management of disasters as they occur. A special emphasis of the UN-SPIDER platform is placed on the use of space communications, remote sensing and meteorological data, mapping, and precision navigation and timing space networks [12].

Nongovernmental Organizations

The global resources to respond to natural disaster include many private resources as well. The most important of these are nearly a hundred nongovernmental organizations (NGOs) that include organizations such as OXFAM, the International Red Cross & Red Crescent, the Catholic Relief Organization, Doctors without Borders, REACT International, Care international, Hope Worldwide, All Hands, International Relief Teams, the Clinton Foundation, the Carter Center, the Bill and Melinda Gates Foundation, and many more [13].

Some NGOs specialize in food and housing, others in medical care, communications service, technical and engineering services, and additional types of aid services. The larger of these organizations liaison with each other and seek some coordination of efforts. Further, some that have witnessed recurring disasters, such as monsoons and the annual floods events in Bangladesh, have begun working on longer-term solutions. In this case, they partner with government officials in the affected areas as well as U.N. agencies such as the International Bank for Reconstruction and Development (IBRD), the World Health Organization (WHO), the Pan American Health Organization (PAHO), and others. Through these efforts, they can create plans and financing for new infrastructure such as dams or levees, or for organizing more permanent resettlement away from areas too close to active volcanic or fault lines.

There are some NGOs, such as the Population Connection Action Fund, that have documented evidence that excessive population growth and increased settlements give rise to disaster events. Their efforts to promote family planning and their publication, *Population Connection*, are designed to address such causes [14].

There are even some NGOS that focus on cosmic hazards. These include the Secure World Foundation, the B612 Foundation, the Planetary Society, the Association of Space Explorers (ASE) (i.e., an association of astronauts and cosmonauts), the International Association for the Advancement of Space Safety (IAASS), and several others. Some are aimed at keeping the general public informed.

In another context, they cooperate with national science organizations and space agencies as well as United Nations agencies. These NGOs primarily focus on space dangers associated with orbital space debris, solar storms, and asteroid, centaur, and comet strikes. In these efforts, they work mainly to influence the policies of national governments, the U.N. General Assembly, the U.N. Security Council, the U.N. Office of Outer Space Affairs, and the Committee on the Peaceful Uses of Outer Space (COPUOS). They support actions to create enhanced safety policies and public programs that address cosmic hazard problems.

Conclusions

The world community is increasingly well organized to support disaster management activities. The United Nations, national disaster relief agencies, space agencies, and military forces can and do support disaster relief efforts. A combination of international organizations, national relief agencies, plus nongovernmental agencies are all pitching in. There is today global disaster-related support for over 200 countries of the world.

That is the good news. The bad news is that there are more natural disasters and calls for relief than ever.

The key to effective disaster management is to be able to provide support at all four stages of preparation, response, recovery, and mitigation. The media, for obvious reasons, tends to focus on disaster response and recovery, but the key to coping with disasters effectively comes most of all at the preparation stage, which includes warnings and alert systems.

In modern times, the nature of disasters has been changed by at least two key factors. First is the drastic impact of climate change, which triggers or intensifies hydrometeorological events, fires, and accompanying problems such as lack of food, mass illness, and lack of access to potable water.

The other factor is when human and natural disasters interact. The impact of a tsunami flooding a nuclear power plant can turn a several billion-dollar disaster into a trillion-dollar disaster. Increasing population growth, urbanization, and settlements in flood plains or near active volcanoes or earthquake-prone regions also can turn what might have been a small disaster into a large one.

The effectiveness of space systems to aid in disaster preparedness has dramatically improved over the years. The information stored in geographic information systems (GIS) is increasing. Thus, the latest information can be displayed on a street-by-street and block-by-block basis to aid recovery workers. Telecommunications satellites can not only provide broadband services during initial response and recovery, but now can also provide ongoing broadband telephonic and data networking services via large-scale LEO constellations, even if terrestrial networks are greatly harmed. The latest remote sensing and meteorological satellite systems can provide detailed and up-to-date information on the extent of damages to buildings, roadways, overpasses, rail lines, utilities, and more. Advances in spatial resolution, imaging rapidity, and new coverage by radar, optical, infrared, and even ultraviolet sensors can provide much greater information about the damaged area. The processing power and AI analytics that now augment the information collection of satellites can provide usable data accompanied by amazing trend analysis. These same space systems can assist with evacuation planning, sending signals to traffic control systems to speed exit from cities.

Big data has now come to the world of disaster management.

This is not to say that all is perfect. Traditional problems, such as a lack of access to local terrestrial electrical power supplies after a disaster, are typically missing. There are problems with those who need to recharge computers or cell and satellite phones, or who need to power medical equipment. Nevertheless, progress continues.

References

1. CRED Crunch. Cost of human disasters. Issue No. 61, December 2020. https://www.cred.be/publications
2. Facts+Statistics: U.S. Catastrophes. https://www.iii.org/fact-statistic/facts-statistics-us-catastrophes. Last accessed 10 Jan 2021
3. EM-DAT international disaster database. https://www.emdat.be/explanatory-notes
4. Scott Madry
5. Pelton, J.N., Madry, S., Camacho-Lara, S.: Handbook of Satellite Application, 2nd edn. Springer Press, Switzerland (2017)., Parts III and IV.
6. Syifa, M., et al.: Landsat images and artificial intelligence techniques used to map volcanic ashfall and pyroclastic materials following the eruption of Mt. Agung, Indonesia. Arab. J. Geosci. **13**(3), 133 (2020)
7. Asia Disaster Preparedness Center. https://www.adpc.net/igo/?. Last accessed 10 Jan 2021
8. Pacific Disaster Center. https://www.pdc.org/. Last accessed 11 Jan 2021
9. Tampere convention: A life-saving treaty. https://www.itu.int/en/ITU-D/Emergency-Telecommunications/Pages/TampereConvention.aspx. Last accessed 12 Jan 2021
10. The disasters charter. http://www.disasterscharter.org. Last accessed 13 Jan 2021
11. General Assembly 46th Session, Resolution 46/182, Strengthening of the coordination of humanitarian emergency assistance of the United Nations: resolution/adopted by the General Assembly. https://digitallibrary.un.org/record/135197?ln=en. Last assessed 16 Jan 2021
12. UN-SPIDER Platform, U.N. Office of Outer Space Affairs. https://www.unoosa.org/oosa/en/ourwork/un-spider/index.html#:~:text=United%20Nations%20Platform%20for%20Space,Emergency%20Response%20(UN%2DSPIDER)&text = UN%2DSPIDER%20aims%20at%20improving, the%20use%20of%20space%20technologies. Last accessed 14 Jan 2021

13. 34 Disaster Relief Organizations. https://www.raptim.org/34-disaster-relief-organizations/
14. Population Connection Action Fund. https://www.populationconnectionaction.org/. Last accessed 15 Jan 2021

13

Artificial Intelligence, Cyberattacks, and Advanced Technology

There is no security against the ultimate development of mechanical consciousness, in the fact of machines possessing little consciousness now. A mollusk has not much consciousness. Reflect upon the extraordinary advance which machines have made during the few hundred years…Assume for the sake of argument that conscious beings have existed for some twenty million years: see what strides machines have made in the last thousand! May not the world last twenty million years longer? If so, what will they not in the end become?

–Samuel Butler, Erewhon, 1871

If a machine can prove indistinguishable from a human, we should award it the respect we would to a human—we should accept that it has a mind.

–Steven Harnad

Introduction

It is hard to accept the idea that a simple string of electrons launched from a single laptop might initiate a nuclear attack on a country, cripple a nation's transit, or poison the water supply of a city of 20 million people. Today, supercomputers, data analytics, and AI are increasingly vital technologies. They are key components of smart cities and efficiently run modern economies. Indeed, these tools have become and will remain crucial for coping with such challenges as climate change, global pollution, or detecting cosmic hazards.

Many people realize that potential cyberattacks are becoming a larger problem, but could this lead to the catastrophic destruction of humankind? Certainly, there have been film fantasies and sci-fi stories about runaway technology. But such fiction is not something to take too seriously, right?

In earlier chapters, the focus was on innovative uses of space systems and new technology for positive purposes. In this chapter, the focus is much the reverse. The objective here is to recognize that such innovative space technology plus data analytics and AI technology could be used in harmful and repressive ways. There is an overwhelming tendency to look to future trends

J. N. Pelton, *Space Systems and Sustainability*, https://doi.org/10.1007/978-3-030-75735-9_13

and evolving technologies as if they will progress in a straightforward, linear fashion—something that this chapter seeks to rectify.

Laws and Risks of Modern Technology

Humans needs to be wary of two very powerful laws. These laws, as defined by Ray Kurzweil among others, have proved very relevant to key patterns of modern technological life.

One is the *Law of Accelerated Returns* (LOAR) (i.e., compound accelerated growth) and the other is the *Law of Unanticipated Consequences* (LOUC). When advanced new technology involving both information technology (IT) and artificial intelligence (i.e., AI) is able to expand exponentially over time, the resulting power of that growth can be enormous. Powerful technologies that might be used to combat certain social, economic, or environmental problems can end up being counterproductive, and more is at risk when that technology falls into the wrong hands. A new vaccine might be modified to become a biological weapon. A new space, information technology, or AI capability developed to identify pollutants or monitor climate change might be used to create a weapons control and targeting system. Machines that "think," monitor, or identify must be studied with a scrupulous eye.

We might reach a time not too far away where smart robotic devices are sophisticated enough to develop a value system, or the right to make life and death decisions over humans. As of the 2020s, the here and now, AI systems are already being developed and employed for war-fighting purposes. One capability that is currently being developed is an AI system that is able to control a fighter jet aircraft in a dog-fight. It turns out that this AI capability is now sufficient to beat and "kill" an ace pilot. The ethical and moral implications of this are significant indeed.

Today, there are also serious concerns about the use of malware by malicious hackers. This might be to rob banks electronically, extort money by means of so-called ransomware, Trojan horses, phishing and pharming schemes, and worse. The much worse attacks would be when cyber-terrorists target governments or vital infrastructure.

When Leon Panetta served as U.S. Secretary of Defense, he warned in 2012 of what he called a future "Cyber-Pearl Harbor" attack on the United States. He urged that the country's strategic defense posture explicitly state that a cyberattack on the United States would indeed be considered an act of war, equivalent to an armed invasion. As it turns out, he should perhaps have mentioned a cyberattack on national elections as well. Secretary Panetta's warning 8 years ago stimulated the United States and many other countries to create a well-staffed and technically competent Cyber Command [1].

Cyber technology can be greatly abused today to attack banks, financial institutions, and vital infrastructure. The recent cyberattack on the United States by Russia was frightening to those who understood its massive scope [2]. The event has become known as the Solar Winds cyberattack. It was initiated by Russian government-funded hackers and sought to infiltrate the U.S. Department of Homeland Security, the U.S. Treasury, and many other agencies, working around the vaunted Einstein 3 governmental cybersecurity [3].

The Einstein 3 cybersecurity system is in place to protect U.S. governmental networks from cyberattacks. Yet while Einstein 3 was adept at finding new, known malware, it was not able to find malware already embedded in the software of commercial suppliers in the chain of supply, such as that supplied by SolarWinds. As a *Washington Post* editorial recently remonstrated: "Agencies ignored a Government Accountability Office report advising them to update a malware-catching tool called Einstein that proved significantly less smart that its namesake." [3] This is only the tip of the iceberg for what is to come in the future.

A small group of expert hackers could attack a nation much more effectively and with much greater impact than an entire army. In the age of supervisory control and data acquisition (SCADA) systems plus IoT-enabled control devices for much of the world's automated infrastructure, these cyber risks are increasingly serious. Yet today, in the age of the "dark web" and sophisticated cyber-crime syndicates, defense is quite difficult. In a future age of smart cities, where an ever-increasing amount of vital infrastructure is programmed for optimized efficiency, the vulnerabilities may well increase. It may be that defense will increase in a linear mode while cyberattack capabilities increase exponentially.

There is a site, known as Check Point "Threat Cloud," that seeks to map global cyberattacks in real time. On the day this was written, December 10, 2020 near midnight, some 37 million such attacks had been mapped around the world [4].

Leaders and Organizations Concerned with AI

Time and again, history teaches us that technological progress tends to follow an exponential rate of growth. Nineteenth-century English novelist Samuel Butler understood this growth curve and the accelerating pace of technological innovation. In his writings (quoted at the start of the chapter), he anticipated the great strides that would come as the industrial age blossomed.

The famous economist Rev. Thomas Malthus, whom we met in previous chapters, also understood compounded exponential growth and feared a future of rapidly increasing humanity, in which there were problems of over-population and lack of food, water, and resources. He did not anticipate, however, that it might be filled with pandemics, super automation, cyberat-tacks, and social media spinning out of control. He most certainly did not worry about a future staffed with cyborg units smarter than those of average human intelligence. Nor did he wonder about whether smart robots might someday begin to debate the usefulness of human existence.

Arthur C. Clarke, whom we have also encountered in this book, had no doubt that artificial intelligence was the greatest invention of the twentieth cen-tury. The question in the twenty-first century is whether the advancement of AI may some day in the not too distant future become a true threat to humanity. Raymond Kurzweil, who invented "Siri" and is the author of *The Singularity is Near*, considers the "Singularity" to be a pivotal point in future human history. This is a time when smart robots attain intelligence equivalent to humans.

Elon Musk has issued several direct warnings on the topic. His most explicit one was made on a YouTube movie back in 2018. In it, he explains:

> If one company or small group of people manages to develop god-like super-intelligence, they could take over the world…At least when there's an evil dicta-tor, that human is going to die. But for an AI, there will be no death—it would live forever. And then you would have an immortal dictator from which we could never escape. [5]

There are many scientists and business leaders around the world who share Musk's concerns. Two leading examples are Jack Dorsey, CEO of Twitter, and the late astrophysicist Stephen Hawking.

One of the lesser-known companies owned by Alphabet—which also owns Google—is a company called DeepMind. It was founded by Shane Legg, Mustafa Suleyman, and Demis Hassabis and is known for its leading role in developing the world's most advanced AI systems.

Stephen Hawking, Elon Musk, and Jack Dorsey, as well as the founders of Deep Mind, issued an open letter together in 2016, citing specific concerns about the development of lethal military AI applications [6]. Despite the con-cerns posed by these famous names, it is very hard to keep the genie sealed up in a bottle. In August 2020, the U.S. Department of Defense released the result of using adapted DeepMind AI software for a military application. Of course, DeepMind had been developed for quite different purposes, but software developers working under contract for the Air Force had then applied it to controlling aircraft engaged in air combat. The following report was given:

> Last week, a technique popularized by DeepMind was adapted to control an autonomous F-16 fighter plane in a Pentagon-funded contest to show off the capabilities of AI systems. In the final stage of the event, a similar algorithm went head-to-head with a real F-16 pilot using a Virtual Reality (VR) headset and simulator controls. The AI pilot won, 5–0 [7].

The Predominant Role of AI and ICT

A Global Trends report from the National Intelligence Council released back in 2008 pointed to many geopolitical shifts predicted to occur by 2025. This report suggested that Asian powers were becoming more economically powerful, with China in particular growing in strength not only economically and strategically, but also in areas such as AI, robotics, smart energy, and other technological fields related to cyber power and military applications. The report emphasized that the combination of AI systems with other technological advances would be crucial to future geopolitical development [8].

Fast forwarding to the latest National Intelligence Council's projection for 2035, the trend lines envisioned 12 years ago are even more strongly forecast. This Global Trends report confirms the basic patterns noted in Chap. 9 on overpopulation: "Automation and artificial intelligence threaten to change industries faster than economies can adjust, potentially displacing workers and limiting the usual route for poor countries to develop." [9]

This new analysis predicted that the application of information & communications technology (ICT), IoT, cybersecurity measures, and AI will not only have an impact on employment, but also on medical and human biological systems, banking and financial systems, political and military capabilities, and industrial efficiencies. The effects of new ICTs on the financial sector in particular are likely to be profound. New financial technologies—including digital currencies, "blockchain" technology for transactions, and AI and big data for predictive analytics—will reshape financial services, with potentially substantial impacts on systemic stability and the security of critical financial infrastructure. This report almost said to be careful of what you wish for, because perceived gains might give rise to social disruption and a lack of resilience over the long run [9].

There is a data analytics organization known as World Data, whose mission is to chart the coverage of different topics as they are reported around the world. World Data thus monitors news reporting of selected media from North America, Europe, Asia, Africa, and South America. It has in fact tracked the frequency of news reporting on the subject of AI, and whether these reports contain a positive or negative perspective.

For North America, the frequency of reporting on AI-related news was highest for CNBC, then descended from there to the *Wall Street Journal*, *Newsweek*, CNET, *Reuters*, CBS News, the *New York Times*, NPR, and so on. Perhaps most significant is that of the 30 media outlets mentioned in the report, only five outlets, including *U.S. Today* and the *San Francisco Chronicle*, covered this technology in a net positive way, and even here, the net plus score was quite low. There were two news outlets that were net neutral, while 23 other outlets had modestly to predominately net negative reports on AI technology.

In contrast, reports in the Chinese press as monitored in the *China Daily* and *Shanghai Daily* were consistently positive by a significant margin [10].

There are significant indications that the Chinese government is using a variety of sensor and AI systems to monitor the behavior of its citizens and then locate and apprehend individuals that have been deemed to have broken the law or engaged in activities considered to be against the best interests of the state. At present, these systems are only able to operate effectively in highly urban areas. In the future, however, high-resolution sensors may be able to monitor the entire countryside of China, allowing full national surveillance. The same of course could be true of many other nations. Indeed, China seems to be exporting such technology for this purpose. An *Atlantic Magazine* article on the subject summarizes this situation as follows:

> Xi's pronouncements on AI have a sinister edge. Artificial intelligence has applications in nearly every human domain, from the instant translation of spoken language to early viral-outbreak detection. But Xi also wants to use AI's awesome analytical powers to push China to the cutting edge of surveillance. He wants to build an all-seeing digital system of social control, patrolled by precog algorithms that identify potential dissenters in real time...And he wants China to achieve AI supremacy by 2030 [11].

Space systems as described earlier can now be used to monitor climate change trends, global pollution, or criminal behavior. But they could also be repurposed for surveillance of citizen activity in what democratic societies would see as sinister ways. Other types of technology, such as that in use for direct broadcasting services, could also be redesigned to monitor the movements of political dissidents and opponents of an entrenched dictatorship. In this instance, the satellite systems, originally used to broadcast television programs in high definition, would now receive signals transmitted from permanently installed anklets or armbands. Such tracking devices would transmit information about the activities and whereabouts of enemies of the state.

This suggests that space systems that have super-resolution capabilities and high temporal resolution (virtually constant coverage of the entire Earth) should perhaps best be owned and operated by nonpartisan and nonpolitical global institutions.

The Ethical Dimensions of AI

Some of the analyses in the Global Trends report looked at how AI and advanced ICT systems might be directly be embedded in humans themselves. The following quote demonstrates the possibility of future developments and hints at related policy and ethical concerns:

> Future retinal eye implants could enable night vision, and neuro-enhancements could provide superior memory recall or speed of thought. Neuro-pharmaceuticals will allow people to maintain concentration for longer periods of time or enhance their learning abilities. Augmented reality systems can provide enhanced experiences of real-world situations. Combined with advances in robotics, avatars could provide feedback in the form of sensors providing touch and smell as well as aural and visual information to the operator. [12]

Artificial augmentation to the human senses, muscular power, or reasoning or memory capabilities is very difficult to halt once such a process has begun. Already, there are pacemakers to control heartbeats, artificial organs, and prosthetics to replace injured or lost legs and arms. Many of these medical procedures are now fully integrated with the human nervous system to allow the prosthetics to respond to the injured person's commands. Science fiction creations such as the Million Dollar Man, the Million Dollar Woman, or Robocop have conditioned public expectation to believe that the integration of computer and AI-based systems with the human brain and nervous and muscular systems can create superhuman capabilities. Such developments could in theory lead to fully integrated computers working with the human brain or artificial sensors implanted in human bodies.

Fig. 13.1 Modified design for U.S. Army robotic "Mule" tank. (Courtesy of the U.S. Army)

Currently, automated warriors do not exist. Even automated or robotic war-fighting systems are limited. Armed drones are remotely controlled by radio relay and human operators. There are prototype robotic tank systems such as the U.S. Army "Mule"; an improved version of the model is shown in Fig. 13.1. Most of these examples are drawn from the United States, since they are most readily available. There are parallel developments most certainly underway in China, Russia, Europe, and perhaps other parts of the world.

These autonomous weapon systems would have the ability to make discretionary life or death decisions. Such capabilities among robotic war-fighting systems, starting with robotic fighter jets, have the potential to create significant escalation in war-fighting missions. On the positive side, enhanced robotic systems with greatly expanded mental and decision-making capabilities could be used to undertake hazardous missions unrelated to warfare, such as mining or deep-sea operations.

It now seems almost inevitable that future AI systems will be programmed with logic systems and heuristic decision-making powers that would ultimately include optimal strategies for taking the lives of enemy combatants. This poses a key moral issue as to whether or not human reasoning must be directly involved. Beyond this issue, there is the haunting concern that robotic machines and software could be turned against wider populations.

How Close Are We to Technology with Human Brain Equivalence?

The goal of some computer system developers, such as Henry Markram and his development team in Switzerland, is to develop a processing and data analytics capability with the "intelligence, heuristic algorithms, processing speeds and memory storage" equivalent to a rat's brain and, perhaps in the future, a human brain.

Markram's "Blue Brain" project actually succeeded in simulating the functional capacity of a rat's brain around 2018. The objective was defined in very precise terms. It set an analytical processing capability goal equivalent to a computer processing speed of over 10^{15} FLOPS and a computer memory of about 5×10^{13} bytes of data. A simulation of what a part of a rat's brain looks like when simulated on a supercomputer is shown in Fig. 13.2.

This work has been described as follows:

The Blue Brain Cell Atlas allows users to visualize all 737 brain regions and the cells they contain, and to download the region with their numbers and loca-

Fig. 13.2 A simulation on a supercomputer of a part of a rat's brain. (Courtesy of the Blue Brain Project and the Swiss Government)

tions. It distinguishes excitatory, inhibitory and some other types of neurons—as well as major types of non-neuronal cells called glia, which insulate and protect neurons. These data are important for researchers trying to understand the structure and function of different brain regions or for building functional models of specific brain regions [12].

When will technological development create an artificial processing and "thinking" capability equivalent to that of the human brain? This would require processing speeds of 10^{18} FLOPS and a memory storage equivalent to 10^{17} bytes of data [13]. This is what Ray Kurzweil has referred to as the so-called "Singularity" breakthrough.

There are number of science fiction writers and movies that have explored themes of AI self-awareness and decision-making. There is Arthur C. Clarke's ominous HAL computer in *2001: A Space Odyssey*. Then there is Isaac Asimov's *I Robot* book, which was significantly reinterpreted and made into a movie starring Will Smith. In that instance, there is an aborted takeover of society by robots and an AI system headed by a master computer called Lisa. Perhaps most grimly, there are the Arnold Schwarzenegger *Terminator* movies, where humans of the future have to fight for their lives against robot-driven extermination. The one flaw in the *Terminator* plot is that humans would likely have little chance in such a war. If an advanced AI capability existed, it would not even have to kill humans directly, as was depicted in the *Terminator* movies. Humanity would be brought to its knees if the intelligent AI simply turned off all the world's smart machines.

The follow-on question is whether such an artificial brain could be taught to learn and reason by some sort combination of embedded software and education system. Would such a machine ultimately develop a personality, have a value system, feel emotions, and become what humans call consciousness? Human nature involves more than the computational elements of memory or processing speeds. Learning, thought, and analytic actions by humans seem to be based on pattern recognition and norms of learned behavior that are not easily quantified or duplicated in digital machine logic systems. Nevertheless, the recent advent of quantum computing, tremendously large memory systems, and advances in heuristics have allowed amazing progress in this direction. The possibility of achieving human brain equivalence may even happen within a decade.

There may well be a future where smart devices become "von Neumann machines," as first defined by John von Neumann of the early twentieth century. It is a mechanism that is not only able to regenerate itself, but also can improve its own designs. This artificial evolutionary development would be much more rapid than biological evolution.

The Challenge of Protecting Modern Society

The issue of all-knowing thinking machines may seem a much grander problem than that of cyber-crime and techno-terrorism, but the power of such individuals in a fully automated world with billions of smart switches and control mechanisms is only growing, as we have already discussed in the context of cyberattacks.

Efforts by companies to reform their data protection capabilities have been mixed. A significant coalition of the major cloud service providers sought universal agreement to what they called an "Open Cloud Manifesto" back in 2014. This was a joint effort by IBM, Cisco, SAP, EMC, and several others, but three of the biggest organizations, namely Google, Amazon Web Services, and Salesforce, declined. The objective of the so-called "Manifesto" was to adopt a consistent set of security protection standards and protocols, plus related auditing and monitoring processes, to ensure the security of their cloud access. Competition to differentiate cloud services by security features is one of the current ongoing problems in the field [14, p. 134].

Banking and credit card security is clearly another global problem. Innovations such as the EMV chip and bank networking and certification systems that have converted to blockchain or two- or even three-factor security systems have brought some improvements for interbank transfers, but there is a host of potential problems still to be repaired. For instance, it was uncovered in 2015 that over 100 banks had revealed vital operating systems information to hackers through phishing schemes. This allowed the ongoing distribution of monies through ATM machines over a time period exceeding 2 years, in an amount that added up to nearly $1 billion [15].

Within the U.S. Department of Homeland Security, there is the Cyber Security and Infrastructure Security Agency (CISA), sometimes known as the Cybersecurity Force. Its mission is to protect Critical Infrastructure and Key Resources (CIKR), in tandem with other U.S. agencies such as the FBI and the National Security Agency. It works with owners and operators of nuclear power plants and research generators, pipelines and electrical grid facilities, transportation systems, and dozens of others to protect CIKR against cyberattacks and other dangers [16].

And other countries have their own protective systems against cyberattacks. These include the Japanese Ministry of Defense Cyber Defense Unit (CDU), the European Agency for Network and Information Security (ENISA), and so on. The most typical approach is to have their cybersecurity and critical infrastructure defense systems located within their ministry of defense or intelligence agency, such as the GRU in Russia [14, pp. 179–199].

The problem for the long run is that associated with the LOAR and the LOUC. There is ever-growing dependence of larger cities on electric grids and supply chains. It was noted in an earlier chapter that the number of megacities will increase from 26 to 41 in a decade. These trends mean a growing number of nations are more vulnerable to cyberattack. Efforts to create more secure networks and information are well and good, but there is the ultimate fact that some people must always know the access codes, and that they could be tricked or bribed into revealing this critical information.

Conclusions

There is special irony in the fact that the latest AI and information and telecommunications technology can make life on planet Earth a much more wonderful place to live, with incredible comforts and ease. Yet at the same time, the automation, cyber networks, remote controls, and other "smart" technologies make modern society much more vulnerable. All of these things could suddenly come crashing to a halt as a result of natural dangers and cyberattacks. Recovery in most instances becomes more difficult as more automation occurs.

Perhaps even more frightening is the idea of an AI-human war as portrayed in the *Terminator* movies. Yet this is in many ways a preposterous image. The AI systems would not have to fight a war—they could simply turn off the automated machinery on which people depend.

For the most part, people are linear in their thinking; technological advances are exponential. The Law of Accelerated Returns (LOAR) and the Law of Unanticipated Consequences (LOUC) in the world of technology can produce changes much more rapidly that people think. There are very good reasons why some of the most farsighted people on the planet have staunchly opposed the use of AI systems to carry out military missions. They have counseled against giving smart machines life and death power over humans. AI-controlled fighter jets are currently under development in the United States, and perhaps other countries are a potential pathway to oblivion.

These are not choices restricted to democratic states. Other countries with totalitarian leadership might race to develop artificial intelligence to carry out cyberattacks on nuclear power plants, banking systems, or biochemical research labs. Once the United States developed atomic weapons, it was only a matter of time before other countries did so. The same is true here. Once lethal artificial intelligence is freed and goes on to "live" on the Internet, the die is cast, the future is shaped, and the whole of humanity should beware.

References

1. Stone, A.: How Leon Panetta's 'Cyber Pearl Harbor' warning shaped the cyber command. Fifth Domain, July 30, 2019. https://www.fifthdomain.com/opinion/2019/07/30/how-leon-panettas-cyber-pearl-harbor-warning-shaped-cyber-command/
2. Sanger, D.E.: Russian hackers broke into federal agencies, U.S. officials suspect. New York Times, updated December 21, 2020. "https://www.nytimes.com/2020/12/13/us/politics/russian-hackers-us-government-treasury-commerce.html
3. Preparing for the next cyberattack. Washington Post, January 6, 2021. p. A22.
4. Live cyber threat map. https://threatmap.checkpoint.com/. Last accessed on 10 Dec 2020
5. Thompson, C.: Elon Musk warns that creation of a 'god-like' AI could doom mankind to an eternity of robot dictatorship. Business Insider, April 6, 2018. https://www.businessinsider.com/elon-musk-says-ai-could-lead-to-robot-dictator-2018-4
6. Finley, K.: AI fighter pilot beats a human pilot, no need to worry (really). Wired Magazine, June 29, 2016. https://www.wired.com/2016/06/ai-fighter-pilot-beats-human-no-need-panic-really/
7. Knight, W.: A dogfight renews concern about AI's lethal potential. Wired Magazine, August 25, 2020. https://www.wired.com/story/dogfight-renews-concerns-ai-lethal-potential/
8. Global trends 2025: A transformed world. National Intelligence Council November 2008. https://www.dni.gov/files/documents/Newsroom/Reports%20and%20Pubs/2025_Global_Trends_Final_Report.pdf
9. Global trends and key implications through 2035. National Intelligence Council. https://www.dni.gov/files/documents/nic/GT-Full-Report.pdf. Last accessed 20 Dec 2020
10. World Data.AI North America. https://worlddata.ai/News?sq=Artificial%2C%20Intelligence&q=NORTH%20AMERICA%20NEWS%20SOURCES&s=NEWS%20AI%2C%20SENTIMENTS&r=GLOBAL%20DATA&sr=WORLDDATA.AI&indices=ZXRsX3YxX25ld3NfbWFzdGVy. World Data.AI China. https://worlddata.ai/News?sq=Artificial%2C%20intelligence&q=CHINA%20NEWS%20SOURCES&s=NEWS%20AI%2C%20SENTIMENTS&r=GLOBAL%20DATA&sr=WORLDDATA.AI&indices=ZXRsX3YxX25ld3NfbWFzdGVy#person. Both URL were last accessed on 7 Dec 2020

11. Anderson, R.: Panopticom is already here. Atlantic, September 2020. https://www.theatlantic.com/magazine/archive/2020/09/china-ai-surveillance/614197/
12. Intelligence report predicts IT in 2035, a world of cyborgs with Asia as top power. https://www.csoonline.com/article/2223664/intelligence-report-predicts-it-in-2030%2D%2Da-world-of-cyborgs-with-asia-as-top-power.html
13. Kurzweil, R.: How to Create a Mind, pp. 124–126. Viking Press, New York (2012)
14. Pelton, J., Singh, I.: Digital Defense. Springer Press, Switzerland (2015)
15. Anderson, M.: Bank-hacking ring may have stolen $1 billion. Washington Post, February 16, 2015. p. A3
16. CISA: Cybersecurity, https://www.cisa.gov/cybersecurity. Last accessed 12 Dec 2020

14

A Global Sustainability Treaty

Humans are burning about 40 gigatons of fossil fuels per year. Scientists have calculated that we can burn about 500 gigatons of fossil fuels before we push the average temperature above 2 degrees Celsius higher…this is as high as we can push it, they calculate, before really dangerous effects will follow for most of Earth's bioregions…
—Kim Stanley Robinson, The Ministry for the Future

Introduction

This chapter was written with extensive contributions from Ram Jakhu and Kiran Mohan Vazhapully.

If the scientific calculations in the above quote are correct, there are only 12.5 years left before planet Earth and human civilization are in serious trouble.

Politicians and business leaders worldwide are often urged to unite and take concrete actions to sustain life and save the Earth. Conceivably, such initiatives would be hard to accomplish without stepping on some very powerful toes. As we have seen, efforts directed at global pollution and climate change are feared to slow economic growth in mainstay programs. The key to progress in part may be to convince politicians that clever and innovative thoughts can also achieve shorter-term gains. Innovative environmental enterprise, better marshaling of existing resources, and seeding of entrepreneurial initiatives might allow sustainability efforts to generate jobs and create new sectors of economic growth.

J. N. Pelton, *Space Systems and Sustainability*, https://doi.org/10.1007/978-3-030-75735-9_14

This chapter sets forth a way for the world community to address the serious challenges that humanity's actions are posing to its own survival. It is about creating a comprehensive arrangement that focuses not on creating a new international organization, but rather on finding better and more cost-effective solutions for collaboration across target areas. The first step forward would be a Comprehensive Agreement for Global Sustainability. If the Panel on Global Sustainability presented here is actually formed and does its job well, the next steps toward a Global Sustainability Treaty[1] should inevitably follow.

A very embryonic draft of this international agreement is outlined at the end of this chapter to demonstrate how such an approach might be structured. As said before, this does not involve creating a new international agency. It does nothing more than propose a global think tank on global challenges, and then kickstart a number of creative processes to help humanity revise its practices.

Rationale for and Approach to Global Sustainability

If the Earth were an apple, the atmosphere would be thinner than the rind of that apple. The point is that the Earth is a delicate and small mechanism when considered in the context of the larger universe. Creating a long-term plan to preserve life on the Earth requires an integrated, universally understood, globally agreed road map accepted by the best scientific minds, politicians across the ideological spectrum, societal and cultural leaders, and the business and economic community.

A piecemeal approach to this issue is imprudent, as global sustainability issues and the major risks defined in this book are inseparable. Climate change, pollution, the proliferation of weapons of mass destruction, population growth, cosmic hazards, the spread of pandemics, growing income inequality within and between countries, and more involve interactive parts that require multidisciplinary thought and analysis. Hence a holistic approach that is considerate of the complexities involved is the need of the hour. This aggregated approach must give due consideration to business interests, economic growth patterns, social and cultural realities, employment needs,

---

[1] This chapter was written by Joseph N. Pelton, Ram Jakhu and Kiran Mohan Vazhapully. The detailed knowledge that Professor Ram Jakhu and Kiran Mohan Vazhapully contributed to this chapter is greatly appreciated.

regulatory processes, and technical knowledge in scientific, engineering, environmental, demographic, and other dimensions. Meaningful incentives to political, business, economic, and societal action may be as important as science and technology for creating sustainable processes.

Further, this process must not get drowned in bureaucracy typified in institutions at all levels. The pressing need is to create a small interdisciplinary team of generalists and experts to agree on several key questions on humanity's survival and durable solutions. They may be drawn from industry, the World Economic Forum, U.N. agencies, governmental ministries, academic institutions, nongovernmental organizations, scientific and cultural communities, and leading thinkers of the world.

Main Questions and Implications

What are the major issues to be addressed? As we have seen, the list includes:

(a) Global pandemics
(b) Pollution, global warming, and consequent increase in natural disasters
(c) The proliferation of weapons of mass destruction (i.e., nuclear and biochemical weapons)
(d) Cosmic hazards (including potentially hazardous solar storms, coronal mass ejections, asteroid strikes, comets, centaurs, and orbital space debris)
(e) Overpopulation and rapid urbanization
(f) Cyberspace and AI threats

In 2020, a Delphi survey was conducted about the current level of perception concerning global existential threats among an international grouping of scientists, educators, and international policy and law experts. The summarized results of the survey are presented in Appendix A.

A careful consideration of global existential threats throws open a multitude of intertwined questions:

- How are these risks and challenges interrelated?
- What kind of programs can address these threats holistically and systematically?
- Are existing overlaps and redundancies, especially between national, regional, and international programs, part of the problem?
- If so, could an interdisciplinary global initiative represent a superior roadmap or plan of action?

- How can complementary initiatives through start-ups and private enterprises, public-private partnerships, competitive challenges, university grants, and the actions of nongovernmental organizations (NGOs) buttress this initiative?
- Could these initiatives also address concerns of economic growth, create employment, and advance scientific knowledge?

Review of Existing Sustainability Capabilities

Some argue that several international organizations already exist to address the most severe threats that humans face, and that the establishment of new institutional structures might lead to duplication of efforts and jurisdictional overlap. Admittedly, groups and mechanisms are now in place to address some of these concerns. However, in some cases, there are serious gaps. Crucial among them is that they miss a purposeful mechanism to understand the bigger picture—how global threats are interrelated and how a multidisciplinary perspective could better detect, articulate, and mitigate the effects.

Here is a quick review of current global capabilities, some of which have been addressed in various ways previously throughout this book.

Global Pandemics

The International Health Regulations (IHR), as revised in 2005, are now in effect for the 196 members of the World Health Organization (WHO). IHR is a formal international arrangement with a binding effect for all WHO members. It lays out a process that allows nations to report health concerns that have the potential to cross borders to the WHO. This involves rapid electronic reporting in accordance with global standards for responding to outbreaks of diseases. State parties to this arrangement are now using a self-assessment annual reporting tool (SPAR) to indicate each nation's capabilities to meet the 13 standards for reporting areas.

The IHR grew out of a global effort to minimize the threat of pandemics. It creates specific reporting obligations related to public health events for countries and grants them rights to effective medical assistance. It also specifies the criteria to be used to determine whether an outbreak of disease, a chemical threat, or a natural catastrophe event constitutes a "public health emergency of international concern." [1]

The Pan American Health Organization (PAHO) works closely with the WHO but focuses on the Americas. Most nations are more positively disposed to support the WHO and follow the IHR because they have a vested interest in combatting the spread of disease and maintaining public health [2].

Most other threat areas are not so clearly defined. That said, one cannot ignore that the IHR and the WHO are suspended in a bubble that does not allow a broader worldview. The system sidesteps rapid systemic changes that are escalating risk factors in the rise of pandemics.

Transboundary Pollution and Climate Change

The leading global institutions that address biodiversity, pollution concerns, and climate change are the United Nations Environmental Programme (UNEP) and the closely related World Conservation Monitoring Centre. Global economic and business interests, population growth, and expansion of urban settlements give rise to a wide range of activities and associated problems. The result tends to be sharp differences in opinion about how significant the risks from pollution are and how developmental needs should be balanced with environmental protection.

As a result, global agreements in these areas are much more challenging to achieve. Previous international agreements include:

- The Antarctic Treaty and its Protocol on Environmental Protection (including Annexes on Environmental Impact Assessment, Conservation of Antarctic Fauna and Flora, Waste Disposal and Waste Management, Prevention of Marine Pollution, Area Protection and Management, and Liability Arising from Environmental Emergencies), the Convention on the Conservation of Antarctic Marine Living Resources (CCAMLR) including its Final Act, and the Convention on the Conservation of Antarctic Seals (CCAS)[2]
- Convention on International Trade in Endangered Species of wild fauna and flora (CITES), 1973 [3].
- Montreal Protocol on Substances that deplete the Ozone Layer (to the Vienna Convention for the Protection of the Ozone Layer), 1987 [4].

---

[2] the Antarctic Treaty and its Protocol on Environmental Protection (including Annexes on Environmental Impact Assessment, Conservation of Antarctic Fauna and Flora, Waste Disposal and Waste Management, Prevention of Marine Pollution, Area Protection and Management, and Liability Arising from Environmental Emergencies), the Convention on the Conservation of Antarctic Marine Living Resources (CCAMLR) including its Final Act, and the Convention on the Conservation of Antarctic Seals (CCAS) (Secretariat of the Antarctic Treaty) https://www.ats.aq/devph/en/news/178 (Last accessed 10 Jan 2021)

- Basel Convention on Transboundary Movement of Hazardous Wastes, 1989 [5].
- Convention on Biological Diversity, 1992 [6].
- United Nations Framework Convention on Climate Change of 1992 [7].
- UN Convention on Desertification, 1994 [8].
- Paris Agreement of 2015 [9].

The World Meteorological Organization (WMO) also plays a key role in the global coordination of satellite systems for meteorological services, monitoring climate change, and observing fundament changes to atmospheric conditions. This is accomplished through the World Weather Watch Programme, the Global Observation System (GOS), and the Coordination Group for Meteorological Satellites (CGMS) [10].

The WMO and UNEP have therefore played a complementary and pivotal role in coordinating global monitoring services of the atmosphere. Their role relates to both short-term changes in weather and atmospheric conditions and longer-term trends. Their coordinative activities involve meteorological satellite systems as well as technical standards related to climate change and pollution tracking. The WMO and the UNEP have collaborated, for instance, on measurement and tracking of such concerns as jet aircraft and rocket launcher air pollution, coastal erosion, desertification, changes to the oceans, and global temperature increases.

Biochemical and Nuclear Weapons

There has been a proliferation of international institutions and regulatory arrangements during the Cold War to regulate the development, use, transfer, and sale of these weapons. They include, among others, the United Nations Security Council and the General Assembly, the United Nations Office of Disarmament Affairs, and the International Atomic Energy Agency. There are many others, including national defense ministries, health agencies, and hundreds of nongovernmental organizations. The number of agreements is quite overwhelming. They include the following:

- The Biological Weapons Convention [11].
- Convention on the prohibition of the development, production, stockpiling and use of chemical weapons and on their destruction, Paris, 1993 [12]
- The Convention on the Prohibition of the Development, Production and Stockpiling of Bacteriological (Biological) and Toxin Weapons and on their Destruction, London, Moscow, and Washington, April 10, 1972 [13]

- Treaty Banning Nuclear Weapon Tests in the Atmosphere, in Outer Space and Under Water, 1963 [14].
- The Comprehensive Nuclear Test Ban Treaty, 1996 [15]
- Treaty on the Non-Proliferation of Nuclear Weapons, 1968 [16].
- Treaty on the Prohibition of Nuclear Weapons of 2017 [17].
- The Treaty between the United States of America and the Russian Federation on Measures for the Further Reduction and Limitation of Strategic Offensive Arms, also known as the New START Treaty, 2011 [18].

Similarly, a multilateral export control regime has been institutionalized to regulate the proliferation of weapons of mass destruction. This regime includes:

- The Wassenaar Arrangement on Export Controls for Conventional Arms and Dual-Use Goods and Technologies [19].
- The Nuclear Suppliers Group for the control of nuclear related technology [20].
- The Australia Group for the control of chemical and biological technology that could be weaponized [21].
- The (Hague) International Code of Conduct against Ballistic Missile Proliferation of 2002 [22].
- The Missile Technology Control Regime for the control of rockets and other aerial vehicles capable of delivering weapons of mass destruction [23]

Cosmic Hazards

This is an area where scientists and a highly specialized group of individuals set policy, decide what is and what is not possible, and control the fate of perhaps billions of people in what might be considered worst-case conditions. The main institutions here are the world's space agencies, agencies concerned with space weather such as the National Oceanic and Atmospheric Agency in the United States, the U.N. Committee on the Peaceful Uses of Outer Space, and few new international institutions, such as the International Asteroid Warning Network, the Space Mission Planning Advisory Group, and nongovernmental organizations such as the Secure World Foundation. Ministries or departments of defense might be consulted, but in most cases, this might come too late. The only international agreement that touches on this issue is the United Nations Outer Space Treaty of 1967 [24].

Overpopulation and Over-Urbanization

Currently, the humanity consumes 1.75 times more than the Earth regenerates in a year. The resources and international agreements that address this issue are likewise limited. U.N. Food and Agricultural Organization has functional expertise, particularly in agricultural research, global demographics, and global demand for agricultural products. The International Bank for Reconstruction and Development also has substantial resources to track global population growth, data related to the economic effects of rapidly growing cities as well as income inequality [25]. The United Nations Secretariat and the United Nations Refugee Organization have limited tools that might assist in this effort to at least measure the severity of the problem. Dozens of institutions can provide some information and relevant data. In this regard, the World Bank's data on income inequality and shared prosperity is quite useful [25].

However, these are almost always unwelcome issues for public policies or national or international agreements to address. The lack of institutional support, regulatory tools, or technologically effective enforcement systems constitute one of the most significant challenges to achieving longer-term global sustainability.

Cyber Warfare

Various committees of the UN General Assembly have deliberated on cybersecurity matters. Issues of cybersecurity have also been witnessed in the UN Security Council in the context of terrorist activity, the UN Economic and Social Council, and various subsidiary organs and specialized agencies. The International Telecommunication Union (ITU) also has some expertise and responsibility in the misuse of technology. However, there is no exclusive international agency nor any particularly relevant international agreement that provides meaningful controls, regulatory tools, or even guidelines for preventing technology abuse in these areas. Here, better tools and agreements will be needed as technology advances at a rapid pace.

Concluding Remarks

As it is evident from the discussion above, global governance has become an important mode to address existential threats. What is also evident is that the existing agreements and regulations are exclusive, and they ignore the interconnectedness of these threats. But all is not lost. The concept of human security, which emerged in the 1994 UNDP Development Report, is on its way to changing the practice of global governance. The ongoing pandemic will give much-needed impetus to this change. Hopefully, such a paradigm shift will make international relations rise above fractious politics.

Humanity must recognize that the modern world is truly at risk, and that the risk is escalating.

There must be a strong commitment to deploy organizational, financial, and intellectual resources enough to succeed in building a sustainable future. A new global sustainability initiative and supporting framework agreement must be devised as a strong foundation to this work. The needed monies and political commitment cannot be a half-hearted a token effort. It cannot afford failures like the League of Nations or some of the earlier environmental initiatives such as the Rio Conference or the Kyoto Protocol.

A global commitment will be pointless and unproductive unless there is a strong agreement by all nations, peoples, and business and political leaders to the following:

- To elevate sustainability and preservation of humanity and natural life to the very top priority
- To commit resources to further longer-term sustainability on Earth and develop viable programs to mitigate severe threats to humanity as needed
- To optimize and consolidate existing institutions and commit multidisciplinary interests as needed to increase capabilities, improve efficiencies, and reduce risks for long-term human survival
- To use public-private institutions, entrepreneurial initiatives, design competitions, nongovernmental organizations, and revamped institutions to achieve sustainability goals
- To assess holistically the existential risks, problems, incentives, and to carry out a SWOT analysis (i.e., Strengths, Weaknesses, Opportunities, and Threats)
- To understand the linkages central to many existential threats

Draft of a Preliminary International Agreement on Global Sustainability

Preamble – Rationale

Whereas there is heightened global awareness of the rising dangers to the world's environment being caused by climate change, the hazards to natural life and vegetation worldwide have grown in recent years, with an alarming increase in the average temperature of the world's atmosphere and adverse other climatic changes related to access to potable water.

Whereas the emergence and spread of a series of coronal viruses have threatened humanity with exponential progression of pandemics in the past decades.

Whereas pollution has become a subject of increasing global concern owing to overexploitation and indiscriminate use of Earth's nonrenewable resources and spur in industrialization across the globe.

Whereas the world continues to see the development and testing of new types of weapons of mass destruction, including nuclear and biochemical weapons, and their delivery systems, despite various attempts to prevent the proliferation and use of such weapons through various treaties, conventions, and transparency and confidence-building measures.

Whereas there have been continued rapid growth of world population, urbanization, income inequality, and large-scale settlements that create more significant target areas for major devastation of megacities by potentially hazardous asteroids, bolides, comets, and centaurs. Likewise, these human settlement areas are also now much more vulnerable to solar storms due to radiation flares and especially coronal mass ejections that could destroy electrical grids, communications and information networks, essential pipelines, satellite networks, and critical control systems such as SCADA and IoT devices.

Whereas there is now an unparalleled increase in the global population, making increasingly large and challenging demands on global resources. Megacities are anticipated to grow in number from 28 in 2014 to 45 by 2030, and cities of 5–10 million are expected to rise over this same period from 43 to 63. They thus will make a global urban population perilously dependent on supply chains that can be disrupted by pandemics, cosmic disasters, natural disasters, and other catastrophes.

Whereas nuclear waste continues to increase at an alarming rate and is threatening the health and longevity of many people and industrial activities, and demand for electrical power continues, as economic systems that are disposable rather than circular and sustainable continue to feed global pollution.

Whereas the natural and manmade disasters that include fire, earthquakes, tsunamis, hurricanes, typhoons, tropical storms, and tornadoes, ocean thermal temperature rises, volcanoes, drought, loss of lakes and aquifers, locust and other insect infestations, red tides, oil spills, deforestation, desertification, and changes to the magnetosphere are becoming more severe due to population growth, income inequality, expansion of the human settlements, pollution of the oceans, and loss of agricultural lands.

Whereas technological advancement, development of automatic weapons, artificial intelligence, and information networks have made global society more vulnerable to cyberattacks.

Whereas all of the potentially existential threats to human civilization and the longer-term sustainability of life on Earth, as noted above, seem to have many interrelated drivers and constitute interdisciplinary areas of scientific, medical, space sciences, economic, political, legal and cultural knowledge, it appears prudent to form an ongoing global sustainability initiative based on a unified and interdisciplinary planning, implementation, and execution process.

The Signatories to this Agreement have agreed as follows:

I. Objective

The objective of this Agreement is to determine and enhance global collaborative efforts to ensure long-term sustainability of human and natural life on Earth.

II. Utilization of Existing Global Resources

The Signatories to this Agreement recognize that integrated, comprehensive, and interdisciplinary planning, implementation, and execution are necessary to mitigate or prevent the risks posed to humanity by global existential threats. They further realize that establishing a new international organization to address these existential concerns could be counterproductive, might limit innovative thoughts and actions, and would create competition among existing agencies, scientific bodies, non-governmental organizations, and business enterprises that might be best able to address these issues if they collaborate.

III. The U.N. General Assembly and Its Sustainable Development Goals

The starting point for addressing those existential risks identified in the Preamble is to review the United Nations Seventeen Sustainable Development Goals (SDGs). The Global Sustainability Initiative participants would review the quantitative measures that record the current rate of progress being made against these various goals. Further, a new goal would be added to the U.N. Sustainable Development Goals. This eighteenth SDG would seek to measure specific progress that might be made to reduce the level of risk to humanity in each area enumerated in the Preamble, and other new risk areas identified by the Global Sustainability Initiative Panel participants. These would be cross-indexed against the progress made in each of the seventeen goal areas. This would allow an annual determination as to where progress in these areas best correlates with the current seventeen goals.

IV. Participants of the Global Sustainability Initiative Panel

The initial participants in the Global Sustainability Initiative Panel of 70 members would be drawn from the following:

The U.N. and Its Specialized Agencies with representation from: the World Health Organization, World Meteorological Organization, the U.N. Environmental Programme, International Atomic Energy Agency, the United Nations Educational, Scientific and Cultural Organization, the International Telecommunication Union, U.N. Office of Disarmament Affairs, the U.N. Food and Agricultural Organization, the U.N. Office of Outer Space Affairs, and The International Bank of Reconstruction and Development (Up to 20 members).

World Economic Forum This includes designated representatives from the World Economic Forum (up to 9 members, each from a different country).

Space Agencies Six representatives from the world's largest space agencies would be designated to participate: National Aeronautical and Space Administration, China National Space Administration, European Space Agency, Japan Aerospace Exploration Agency, Roscosmos, and Indian Space Research Organisation (up to 6 members).

Nongovernmental Organizations Ten experts from representatives drawn from world-recognized, leading NGOs would be designated by the Secretary

General of the UN. They would be selected from representative organizations from around the world that work in environmental studies, disarmament and peace-keeping, astronomical hazards, medical research, meteorological studies, population studies, urban studies, and legal affairs (up to 10 members).

Leading Thinkers of the World They would be designated by the U.N. General Assembly. They might be Nobel Prize winners, world-renowned academics, noted jurists, leading journalists, and business leaders from around the world (up to 15 members).

Representatives from the Developing World All nations of the United Nations with gross domestic product per capita under $2000 per year will be invited to nominate representatives to the panel. Ten representatives would be selected by lottery from these nominees (up to 10 members).

V. Functioning of Global Sustainability Initiative Panel

This Panel would meet as needed for the first year and then once a year to review the progress made on identifying various global existential risks as displayed on a web-based "Global Threat and Sustainability" display. They would also receive a preliminary review in designated sub-panels and then as a panel-of-the-whole. This international and interdisciplinary Panel would review progress made toward new and improved actions to mitigate and lower the level of these risks. The purpose of the interdisciplinary review and discussion would be to understand linkages between various threats.

The first meeting of this Panel would be in person and be convened at the Vienna International Centre, Vienna, Austria. Subsequent meetings may be held electronically. The primary purpose of the first meeting is to carry out two mandates. The first mandate would be to agree on a specific list of global potential existential risks that could include up to 20 existential threats to human and natural life on Earth. All of the global risks included in the approved listing would provide (i) a brief description of the global threat, (ii) major contributing factors to the risk, and (iii) potential actions, programs, technologies, or systems that might be developed, perfected, or implemented to address these risks.

Secondly, this group would agree on an interdisciplinary staffing plan of no more than 75 professionals and 25 support staff. Under their elected head, this staff would work on addressing these planetary risk factors and mitigating strategies, risk reduction programs, competitive challenges, research programs, panel discussions, new start-up ventures, or other initiatives during the course of the year. This staffing plan would define the primary location of the Initiative. However, there would be a broad understanding that most participants could work from their home offices and participate electronically in the efforts of the Initiative.

The seconded staff, even though working remotely, would dedicate 100% of their worktime for the Initiative. However, exceptions could be agreed upon under special conditions. The Panel would also elect an executive to lead this global effort. Panel will prepare documents and deliberative processes and indicate: (i) key potential existential threats and their initial description in terms of why they constitute major risks, along with ways that these threats might be mitigated; and (ii) possible key professionals and support staff.

VI. Meetings of the Global Sustainability Initiative Panel

Subsequent annual meetings of the Panel will be held electronically to review annual reports on progress made to mitigate these risks, indicate the correlation between progress on the U.N. Sustainable Development Goals, assess the progress made toward monitoring each risk factor, progress as regards formulating programs or strategies to mitigate or contain the identified threat, and recommend changes to the threat list. Such meetings would develop a final report that would go to the U.N. General Assembly, the Heads of State of all nations in the world, and all the participants in the Global Sustainability Initiative Panel. It would identify programs and initiatives with an indication of how these efforts might be undertaken.

VII. United Nations Global Sustainability Risk Assessment Process

The United Nations Secretariat and other entities designated by the Secretary-General of the United Nations, in cooperation with the Panel, shall prepare each year, subsequent to the first report of the Global Sustainability Initiative Panel, a public report on all of the risk factors identified by the Panel as existential threats to human sustainability and natural life on Earth. This report will provide an assessment of the current level of efforts to detect these various threats, assess the level and probability of their occurrence, and assess the degree of effectiveness of global cooperative efforts to mitigate them. This report shall seek to achieve a common position on these matters and shall be published, distributed online, and released worldwide. There would be a Global Risk Alert issued to the world press that indicates each risk area. On a ten-point color-coded scale, the currently perceived level of risk to the continued sustainability of human civilization will be indicated for each and every risk area. This Global Risk Alert will indicate progress or risk increase for each threat concern agreed by the Panel.

VIII. Fifth-Year Threshold Review of Progress

The fifth Meeting of the Panel would formally review the progress made on each of the identified planetary risks to understand better the degree of risk associated with each identified threat. The Panel would also review (i) the degree to which progress had been made to mitigate each risk, (ii) the results of research undertaken, development efforts, challenge competitions, new entrepreneurial initiatives, and conferences and panel discussions held. The result of this review would be submitted in the form of a report to the U.N. General Assembly, the Heads of State of all nations in the world and all the participants in the Global Sustainability Initiative Panel. Such a report would include the case for and against the continuation of this Global Sustainability effort.

IX. Continuing Effort for Global Sustainability

The work of the Panel and the expert staff shall continue after the first 5 years unless the Global Sustainability Panel decides the work as mandated in this agreement has achieved its purpose or is not of sufficient value and productivity to continue.

References

1. International Health Regulations of 2015. https://www.who.int/health-topics/international-health-regulations#tab=tab_1
2. https://www.who.int/transplantation/mission/en/
3. https://cites.org/eng/disc/text.php
4. https://ozone.unep.org/treaties/montreal-protocol/montreal-protocol-substances-deplete-ozone-layer
5. http://www.basel.int/TheConvention/Overview/TextoftheConvention/tabid/1275/Default.aspx
6. Convention on Biological Diversity, 1992. https://www.cbd.int/doc/legal/cbd-en.pdf
7. United Nations Framework Convention on Climate Change of 1992. https://unfccc.int/resource/docs/convkp/conveng.pdf. Last accessed 20 Jan 2021
8. https://www.unccd.int/sites/default/files/relevant-links/2017-01/UNCCD_Convention_ENG_0.pdf
9. The Paris Accords. https://unfccc.int/process-and-meetings/the-paris-agreement/the-paris-agreement. Last accessed 20 Jan 2021

10. Camacho-Lara, S., Madry, S., Pelton, J.: International meteorological satellite systems. In: Handbook of Satellite Applications, 2nd edn, p. 1097ff. Springer Press, Switzerland (2018)
11. The Biological Weapons Convention. https://www.state.gov/about-the-biological-weapons-convention/. Last accessed 10 Jan 2021
12. Convention on the prohibition of the development, production, stockpiling and use of chemical weapons and on their destruction, Paris, 1993. https://ihl-databases.icrc.org/applic/ihl/ihl.nsf/INTRO/553
13. Convention on the Prohibition of the Development, Production and Stockpiling of Bacteriological (Biological) and Toxin Weapons and on their Destruction, done at London, Moscow, and Washington, April 10, 1972. https://www.state.gov/biological-weapons-convention-text/. Last Accessed 9 Jan 2020
14. Treaty Banning Nuclear Weapon Tests in the Atmosphere, in Outer Space and Under Water, 1963. https://treaties.un.org/pages/showDetails.aspx?objid=08000002801313d9
15. The Comprehensive Nuclear Test Ban Treaty, 1996. https://www.ctbto.org/the-treaty/treaty-text/
16. Treaty on the Non-Proliferation of Nuclear Weapons, 1968. https://www.un.org/disarmament/wmd/nuclear/npt/text
17. https://treaties.un.org/doc/Treaties/2017/07/20170707%2003-42%20PM/Ch_XXVI_9.pdf
18. The Treaty between the United States of America and the Russian Federation on Measures for the Further Reduction and Limitation of Strategic Offensive Arms also known as the New START Treaty. 2011. https://www.state.gov/new-start/
19. https://www.wassenaar.org/
20. https://www.nuclearsuppliersgroup.org/en/
21. https://www.dfat.gov.au/publications/minisite/theaustraliagroupnet/site/en/index.html
22. https://www.hcoc.at/?tab=background_documents&page=text_of_the_hcoc
23. https://mtcr.info/
24. Treaty on Principles Governing the Activities of States in the Exploration and Use of Outer Space, including the Moon and Other Celestial Bodies (The Outer Space Treaty of 1967). https://www.unoosa.org/oosa/en/ourwork/spacelaw/treaties/introouterspacetreaty.html
25. https://www.worldbank.org/en/topic/isp

15

A New Way Forward

It is better to avert a disaster than to be caught by it.
–Anonymous

Oh what a tangled web we weave...
–Sir Walter Scott, from Marmion: A Tale of Flodden Field
There is always a plausible solution to every human problem—neat, easy and wrong!
–H.L. Mencken

Existential threats to humanity are often most easily pushed to the bottom of the priority list. This is because they are difficult and expensive to address, seemingly far away, and most of all people just do not like bad news. A dent in one's car is something to accept. The end of human civilization is something to deny. But evidence is mounting that our activities are creating a web of complex problems that are increasingly dangerous to our long-term survival.

The time has come to marshal the world's best minds and technology in a collaborative effort to cope with the problems discussed in this book. The time has come to start preparations for defense against known global risks. The time has come for humans to demonstrate that they are truly smarter than the dinosaurs—or maybe not.

Response and recovery are only two steps in a four-step process to cope with disasters. Natural disasters must be dealt with through Preparation, Response, Recovery, and Mitigation. Preparation is always the most important factor, but legislators are willing only to make large appropriations to address a problem only after a disaster has occurred. Today, the sad fact is that most preparations for disasters are underfunded and not as well coordinated as they should be. In terms of global pandemics, the World Health Organization collects information and declares national pandemics, but it is not empowered to control many of the critical decisions on healthcare practices, quarantines, vaccine development, or distribution of global medical supplies. The situation with global pollution is perhaps even worse. Neither the

J. N. Pelton, *Space Systems and Sustainability*, https://doi.org/10.1007/978-3-030-75735-9_15

U.N. Environmental Programme, the World Meteorological Organization, nor the Food and Agricultural Organization (FAO) have what might be called actual controls over pollution of the oceans, seas, lakes or rivers, farmlands, the atmosphere, the land masses, or the icecaps. Pollution and pollution controls represent global issues, but nation-states like to pretend that they can control pollution acting alone.

Ocean-related problems such as plastic poisoning, acidity, and the warming of the oceans, which are killing off coral reefs, the algae, plankton, and cyanobacteria that produce most of the planet's oxygen, are global problems that national legislation cannot adequately control. There are a few treaties related to the Law of the Seas, but these have significant holes. Airplanes and rocket launchers can expel exhaust into the atmosphere at will. Ships can dump waste into the oceans, and industries have wide scope to expel pollutants in different countries.

Regulatory authority, the power of the purse, and punitive law enforcement powers are concentrated in the hands of national environmental protection authorities. But the fundamental nature of the problem is global, and so national regulation is inadequate. If continued, human activities in the Anthropocene Epoch will become lethal to life on Earth. Something needs to change.

This period is called the Anthropomorphic Epoch because geologists from around the world have agreed that human actions are the prime shapers of the world's geology, atmosphere, oceans, and biosystems. This is an epoch in which human-driven behavior and vulnerabilities now mutually interact with and reinforce each other—and not in a good way. The results are largely compounding a number of opposing or destructive forces. Examples include how the Fukushima tsunami led to a massive nuclear disaster, or how the California droughts plus human development into California deserts are now resulting in a massive onslaught of uncontrolled forest fire. The examples go on and on.

It is a mistake to assume that the most recent types of threats that have occurred and with which people are most familiar will be representative of the next major crisis. In Appendix A, the questionnaire group rated natural disasters as the most dangerous of the threats to be faced. This has been true in the past. Yet a major asteroid or comet strike or a high-energy solar storm could prove much more dangers and take many more lives. The focus is very often on the dangerous we have known and experienced recently, rather than possible danger that has not occurred for a very long time. The time has come for the world's space agencies to define planetary defense against cosmic hazards as a top-tier strategic objective. In this new era, space agencies would work together to defend planet Earth against solar storms, magnetic polar shifts, potentially hazardous asteroids, comets and centaurs, orbital space debris, and supernovae radiation affects.

Virtually every one of the dozen or so planetary threats that humanity faces today are heightened by rapid population growth and intensive urbanization. There is also a correlation between economic systems that are premised on continuous expansion and throw-away products. A sustainable world is closely linked to a much more sustainable economic system based on recyclable, clean, long-lasting products and services.

The first mistake in seeking human sustainability and survivability is dividing and distributing tasks among as many national actors as possible. The second mistake is overspecialization, thereby obscuring how these various tasks directly relate to each other. The third key mistake is removing any type of unified controls or systemic coordination at the global level.

The European Launcher Development Organization (ELDO) is a quite telling example. This agency sought to develop a European multi-stage rocket. The decision was taken to divide the creation of the multistage rocket, in a nation-based, programmatic way. Thus, a different country was set the task of developing a different part of the new rocket. The various stages worked successfully at least once, but they never performed successfully altogether. In the end, the ELDO "Blue Streak" launcher never achieved a single successful launch to orbit. As a consequence, the European Space Agency was eventually formed as an integrated organization, and it has achieved much greater success in accomplishing its mission (Fig. 15.1).

The moral of this story is that any attempt to parcel out assignments when creating an integrated product is not a formula for success. One has only to look at the recent global failure to control the Covid-19 pandemic to understand that countries need to find a way to work more cooperatively together on common planetary goals.

Fortunately, there are elements of needed and indeed fundamental change contained within the U.N. General Assembly's Seventeen Sustainable Development Goals. Key progress, for instance, has already been achieved with regard to international mechanisms on asteroid safety. There has been broad agreement regarding the need for the International Asteroid Warning Network (IAWN), the Impact Disaster Advisory Group, and the Space Mission Planning Advisory Group. Also, the 21 recommendations of the U.N. COPUOS that came from the Working Group of the Long-Term Sustainability of Outer Space Activities have found general agreement. Then, there are specialized agencies that are also working on some aspects of these sustainability issues: the U.N. Environmental Programme, the Food and Agricultural Organization, the World Meteorological Organization, the

Fig. 15.1 The ELDO Blue Streak Rocket on display at the Deutsch Museum. (Courtesy of the Deutsch Museum)

World Health Organization, the World Bank (i.e., the IBRD), and the U.N. Committee on the Peaceful Uses of Outer Space.

There are also a number of relevant international treaties. These include: the UN Convention of the Seas, the UN Framework Convention on Climate Change, the Paris Agreement of 2015, the Antarctic Treaty and its Protocol on Environmental Protection and its 7 Annexes, the U.N. Outer Space Treaty and its 4 derivative Conventions and International Agreements, the various disarmament agreements related to weapons of mass destruction, and the many other agreements noted throughout this book.

No one would argue that the United Nations is a perfect organization, nor that its specialized agencies are a model of efficiency. Yet, it is the only effective instrument of global policymaking and concerted action among the countries of the world that is available on this planet today. It has succeeded since the World War II on keeping the major nations from attacking each other and also preventing the use of nuclear weapons in wars. Its efforts to create global goals for sustainability at least give some grounds for hope (Fig. 15.2).

The United Nations' Millennium Development Goals and the current Sustainable Development Goals for 2030 are useful starting points to build

Fig. 15.2   The United Nations Organization Building in New York City. (Courtesy of the Wikimedia commons)

toward a program of action that incorporates key elements of the Paris Accords. Such improvements are at the heart of the proposal to create a new International Agreement that could ultimately lead to a Global Sustainability Treaty (GST). The starting place is to create a global panel of experts to agree on a primary list of existential threats to human's longer-term survival. The next step would be to proceed forward with a small staff, which would be seconded from existing agencies and institutions; this team of global collaborators could create within a few years' time a clear presentation of the forces, trends, and regulatory processes that could assist in bringing these various threats under better control.

Key to any successful new goals are clear measurement tools. The vision for the future must have tangible objectives that can be measured quantitatively. They must also address some often taboo subjects that politician seek to avoid. This includes topics such as population growth, evolution towards a recyclable and circular economy, and the impact of super-automation and artificial intelligence on human employment.

There is hope in the fact that satellite networks for telecommunications, data networking, remote sensing, weather and climate change monitoring, and precise navigation and timing services are now global in their physical configuration, scope of operation, frequency of updates, and ability to identify adverse changes via data analytics. This technology can provide more accurate information regarding specific threats and their causes, helping to create a more sustainable and safer world.

It is the thesis of this book that humans now have the knowledge and technology to create a sustainable world that can save civilization for the longer term. Many colleagues from the space community feel that the only hope for the future is to create sustainable places for humans to live in deep space— perhaps on the Moon, Mars, or elsewhere in the solar system. Yet logic, economics, and technological progress, plus the visionary views of key people of vision, from Sir Arthur C. Clarke to James Lovelock suggest that Earth must remain Plan A, Plan B, and Plan C for humanity at least for centuries to come. An International Agreement that ultimately leads to a true Global Sustainability Treaty is a critical and needed step forward.

Appendix A: International Survey About Possible Existential Threats to Earth

During 2020, as a part of the research for this book, a modified Delphi Survey was conducted among an international grouping of scientists, international policy and legal experts, and others. Here are the summarized results.

The first survey question was about the potential types of existential threats that might act as an impetus to form a more coherent global organization.

The result was that the strongest likely forces that might push the world community to create a planetary defense capability were a tie between "climate change" and "efforts to protect against a known cosmic threat," such as a potentially hazardous asteroid or comet. This was followed by "global pandemic," "global war," and "overpopulation."

Overall, there was no strong indication that the world would quickly move in this direction. On a scale rated from one to ten (i.e., from low to high), no compelling impetus to organize globally for this purpose was apparent. Both "climate change" and "defense against cosmic hazards" averaged 6.6. This was followed by "global pandemic" at 5.7, "global warfare" at 5.2, and finally, "massive global overpopulation" (and presumably large-scale starvation issues), which came out at a relatively low 4.1. The apparent view was that global overpopulation and subsequent starvation were something that happens, and a worldwide planetary defense organization could not cure such a problem in any event.

This was followed by a question as to when such a global organization for planetary defense might be created. Forty percent said by 2050. Some 22% optimistically said 2030, 23% said 2075, and 11% said never. If analyzed

J. N. Pelton, *Space Systems and Sustainability*, https://doi.org/10.1007/978-3-030-75735-9

using a bell-curve, then 2050, or 30 years from now, might be considered the most realistic overall response.

Further, they were asked whether such a coordinative body would be formed within the U.N. organization, or perhaps through a new type of organizational structure or process. Only 17% said yes, 45% said maybe, and 38% said some new type of structural organization, such as a public-private partnership, an economic and geopolitical alliance, a regional alliance, or an outgrowth of nongovernmental organizations that work in the disaster relief area, etc.

Another question was about the type of existential threats that could most threaten the existence of humanity for the longer term, primarily in terms of their likelihood of occurrence and deadliness of impact. It seems likely that some respondents considered the likelihood of occurrence in their responses more heavily, while others considered the deadliness of impact as the prime issue. Table A.1 shows the results in descending order. Again, the scale is 1–10 and the "rating" represents the cumulative average response.

Finally, there were many meaningful comments received in the survey that are provided here in full or in part.

- The COVID-19 Pandemic has not been handled well globally. And this can be said about nearly all existential threats to humankind. One gets the

Table A.1 Which threats today endanger human civilization the Most

Rating (1–10)	Type of disaster
8.3	All type of natural disasters including floods, violent storms, tsunamis, earthquakes, volcanoes and super volcanoes, etc.
6.9	Climate change
5.8	Pandemics
5.75	Biological-chemical and nuclear weapons
5.6	Global warfare
5.2	Global food and water shortage
5.1	Air and water pollution
5.1	Asteroid or other cosmic hazard
4.8	Solar storm/coronal mass ejections (Carrington Event or Maunder Minimum)
4.4	Loss of Earth's natural protection against cosmic hazards (ozone layer, magnetosphere)
3.5	Orbital space debris

Notes: Others added concerns related to mass migrations, super volcanoes, and runaway artificial intelligence. These items were not on the original survey list uneasy feeling that these hard to visualize threats are being handled by underfunded experts who operate in small offices and labs, more in the spirit of hobbyists. It is time to professionalize, coordinate, and scale these efforts. The greatest frustration is that at the end of the day, nobody is responsible for defending earth and humanity. Quite literally, our lives depend on major reforms in this area. (Michael Potter, Paradigm Ventures and Co-Founder of Geeks without Frontiers)

- It is becoming increasingly apparent that humanity needs to pay more attention to the fragile nature of the global systems, both natural and artificial, on which it depends. Their interconnectedness means that an impact on any one can have an effect on many of them, leading to the possibility of near-existential damage to our planet, our environments, our fellow creatures and ourselves. (Prof. Chris Welch, International Space University, France)
- Global problems ought to be solved at global level with global participation. An appropriate multidisciplinary approach is indispensable to address multifaceted issues. (Prof. Ram S. Jakhu, Acting Director, Institute of Air and Space Law, McGill University, Canada)
- I am convinced that it is vital for us to begin to take global catastrophes more seriously in the context of both space technologies and international treaties and organizations. We have seen the damage done by Covid-19 by our lack of preparation and our unwillingness to address low likelihood / high-risk events. We must not allow this to happen again. Now is the time to begin to address and prepare for the next crises. (Prof. Scott Madry, University of North Carolina, USA)
- At the present time, the international community has shown little interest in joint-projects to solve global issues that affect us all. (Matteo Madi, Ph.D., Founder of Sirin Orbital Systems, AG, Zurich, Switzerland)
- With swift and painful force, 2020 has revealed how our species is fundamentally interconnected. To the pessimist, this interconnectedness is fragility. To the optimist, however, our interconnectedness is a great strength. And while present global circumstances may look bleak, remember that an optimist always finds multiple unique and novel solutions to problems that the pessimist sees as incurable. It is a strength because (for the first time in human history) we have the global means to defeat the forces of existential risk that have forever threatened life here on Earth. People around the world are waking up to the reality that collective threats with severe risks to our way of life require collective, international coordination and action, and not just by governments when the get around to it. Seeing the way forward is the first step. (Chris Johnson, Secure World Foundation, USA)

- The knowledge and understanding we gain of our own world by going into space has been, without a question, the most profound change in world view ever experienced by the human race. It has opened our eyes. We now know that the Earth is orbiting our sun at a hazardous location where asteroids cross our path just waiting to slam into us and where we are bathed in lethal radiation that seems to change unpredictably. Our planet can and will claim our lives, as a reaction to our way of life and because it is evolving in its own way that we are just now becoming aware of. With this new knowledge we must continue to gain the ability to shape our own future evolving from living on a planet to living on many planets and achieving a new stature within our galaxy. It is truly the most exciting time to be alive. (James Green, Chief Scientist, NASA, USA)

- The most likely threat event to occur is unlikely to be the next one to actually occur. This means that the very low probability but high consequence events are very difficult to predict. (Darren McKnight, Centauri Corporation, USA)

- We are a forgetful species. Why do we forget past events and periods that have decimated our species and laid waste to our fertile lands, time and again? Perhaps we are wired this way to avoid paralysis through analysis? Cycles of drought and famine, extreme climate, and more have all happened repeatedly in the past too. Certain aspects of what you mention could be done by a World Government of sorts that does not impinge on cultural diversity. (Prof. Madhu Thangavelu, Viterbi School of Engineering, University of Southern California, USA)

- The picture of our blue planet, taken by Apollo Astronauts is one of the greatest achievements of the last century. It shows that we are all astronauts on this spaceship and responsible for its vitality. (Prof. Giovanni Fazio, Harvard University, USA)

- We find ourselves somewhat at a crossroad. There is much at stake here and we must avoid a "tragedy of the commons" type scenario. We need to develop a regulatory and cooperative framework for broad and effective measures to ensure that the Earth can be protected from major natural and human-created disasters. Such an effort should encourage and support cooperative actions that are—to borrow a phrase from the Outer Space Treaty—clearly "in the interest of all countries" and represent the "province of humankind." (Prof. Steven Freeland, University of Western Sydney, Australia)

- Many catastrophic events can drastically affect livelihoods and quality of life and also result in widespread loss of life. We should examine these threats more thoroughly and involve younger people in these efforts. This would be one way for the Next Generation of young people around the world to be informed and become leaders in this field. In due course, it will be they who will establish the necessary global cooperative/collaborative organizations to meet these threats. (Gary Martin, International Space University, France)

Index

© The Author(s), under exclusive license to Springer Nature Switzerland AG 2021
J. N. Pelton, *Space Systems and Sustainability*, https://doi.org/10.1007/978-3-030-75735-9